T0353465

Ring
and
Field
Theory

Ring
and
Field
Theory

Kaiming Zhao

Wilfrid Laurier University, Canada

NEW JERSEY · LONDON · SINGAPORE · BEIJING · SHANGHAI · HONG KONG · TAIPEI · CHENNAI · TOKYO

Published by

World Scientific Publishing Co. Pte. Ltd.

5 Toh Tuck Link, Singapore 596224

USA office: 27 Warren Street, Suite 401-402, Hackensack, NJ 07601

UK office: 57 Shelton Street, Covent Garden, London WC2H 9HE

British Library Cataloguing-in-Publication Data
A catalogue record for this book is available from the British Library.

RING AND FIELD THEORY

ISBN 978-981-125-577-9 (hardcover)
ISBN 978-981-125-578-6 (ebook for institutions)
ISBN 978-981-125-579-3 (ebook for individuals)

For any available supplementary material, please visit
https://www.worldscientific.com/worldscibooks/10.1142/12819#t=suppl

Printed in Singapore

PREFACE

Topics included: rings, subrings, quotient rings and ring homomorphisms; field of quotients of an integral domain, ideal theory, isomorphism theorems; unique factorization domains, principal ideal domains, Euclidean domains and Gaussian integers; polynomial rings over unique factorization domains, Schönemann-Eisenstein Irreducibility Criterion for unique factorization domains, Perron's Criterion, Cohn's Criterion and Osada's Criterion; Noetherian rings, modules over rings, free modules, finitely generated modules over Euclidean domains, Smith normal form of matrices over Euclidean domains; fields, field extensions, algebraic closure, finite fields; splitting extension fields, separable extension fields, perfect fields, finite normal extension fields; the Fundamental Theorem of Galois Theory and solvability by radicals.

Overview and approach: In addition to being an important branch of mathematics in its own right, ring and field theory (Galois theory) is now an essential tool in number theory, geometry, topology, Lie groups, algebraic geometry, differential equations.

The main object of study in Galois theory are roots of single variable polynomials. Many ancient mathematics civilizations (Babylonian, Egyptian, Greek, Chinese, Indian, Persian) knew how to solve quadratic equations. Today, most middle school students memorize the "quadratic formula" by heart. While various incomplete methods for solving cubic equations were developed in the ancient world, a general "cubic formula" (as well as a "quartic formula") was not known until the 16th century Italian school. It was proven by Ruffini and Abel in 1824, that the roots of the general quintic polynomial could not be solvable in terms of nested roots. Galois theory provides a satisfactory explanation for this. More generally, Galois theory is all about symmetries of the roots of polynomials. An essential concept is the field extension generated by the roots of a polynomial, called the splitting field of a polynomial. The philosophy of Galois theory has also impacted other branches of higher mathematics (Lie groups, topology, number theory, algebraic geometry, differential equations). This book will provide a rigorous proof-based modern treatment of the main results of ring and field theory (Galois theory).

About this book: This book was originally intended as a textbook for a one-term senior undergraduate (or graduate) course in ring and field theory, or Galois theory. The students are required to have some knowledge on calculus, linear systems, determinants and matrices, and to have taken a first course on abstract algebra. This book can also serve as a reference for professional mathematicians. Earlier drafts of this book were used several times when the author taught MA475 (and MA675), the third course in abstract algebra at Wilfrid Laurier University. When the author prepared these lecture notes he mainly took [F] as his reference.

The author tries to make this book self-contained. Readers will be only required to have some knowledge of first and second year university math background. The book contains 241 carefully selected exercise questions of varying difficulty which will allow students to practice their own computational and proof-writing skills. Sample solutions to some exercise questions are provided, from which students can learn to approach and write their own solutions and proofs. Besides standard ones, some of the exercises are new and very interesting. Some are rather hard. It is not a surprise if the reader cannot solve some of the exercises, particularly for first learners.

Feature of this book: The book is written in a way that is easy to understand, simple and concise with simple historic remarks to show the beauty of algebraic results and algebraic methods. This makes the book a small one which can help students build up their confidence so that they can easily pick up big volumes to learn in their future academic career. The book provides a lot of interesting examples that illustrate definitions, theorems, and methods, and help students to learn how to approach and write correct and good proofs.

A guide for the instructor: This book was originally intended as a textbook for a one-term (36 lectures of 50-minute, or 24 lectures of 80-minute) senior undergraduate or graduate algebra course. For a one-term course an instructor can cover Chapters 1–4 in the order of the book, or can cover Chapters 1, 2, 4, 5 and 6 in the order of the book and consider some sections and some theorems as optional material (for example, Sections 1.8, 2.5, 6.5 and 6.6; the proof of Theorem 4.3.11; Theorems 4.3.13, 4.4.9, 4.4.10, 5.4.16 and 6.2.1; the second half of Sections 5.4 and 6.4). For more in-class examples an instructor can take some exercise questions with solutions from the book. Of course, these are just guides, and an instructor will certainly want to customize the materials in the book to fit his/her own interests and requirements.

A guide for the student:

(1). Always attend to lectures. Class provides informal discussions, and you will profit from comments of your classmates, as well as gain confidence by providing your insights and interpretations of a topic. Don't be absent!

(2). Ask and answer questions during the lecture: Is it correct? I do not get it. Can you explain it?

(3). Take a fresh look at the notes on the topic once the lecture of a particular topic has been given. Well understand definitions, theorems, corollaries and examples, and try to memorize them.

(4). Try to solve more exercises questions in the textbook. Write detailed full solutions to more exercise questions in your own words.

(5). Don't fall behind! The sequence of topics is closely interrelated, with one topic leading to another.

(6). Work in study groups. Have weekly meeting with your group! Articulating your ideas to others will ensure you know materials better and gain confidence.

(7). When midterm exam or assignments are returned, rework the problems on which you lost points to find out exactly what you did wrong. Then carefully read and understand all solutions.

Feel free to contact the author at kzhao@wlu.ca for any questions involving the book (e.g., comments, suggestions, corrections, etc.). The author welcomes your input.

Acknowledgments: The author would like to thank Dr. Dongfang Gao for pointing out some errors in the early draft of the book.

At last, the author wishes all instructors and students who use this book a happy mathematical journey they will undertake into this delightful and beautiful realm of algebra.

Kaiming Zhao
Department of Mathematics
Wilfrid Laurier University
75 University Ave. W., Waterloo
Ontario, Canada, N2L 3C5

June 1, 2021

Contents

NOTATIONS

The readers are supposed to be familiar with the following conventional notations throughout this book:

- $|A|$: the cardinality of the set A,
- $B^A = \{\text{all maps } A \to B\}$ where A and B are sets,
- $\emptyset$: the empty set,
- $a \in A$: a is an element of the set A,
- $A \subseteq B$: the set A is a subset of the set B,
- $A \subset B$: the set A is a proper subset of the set B,
- $A \cup B$: the union of the sets A and B,
- $A \cap B$: the intersection of the sets A and B,
- $A \setminus B = \{a \in A : a \notin B\}$ where A and B are sets,
- $\mathbb{Z}$: the set of integers,
- $\mathbb{N}$: the set of positive integers,
- $\mathbb{Q}$: the set of rational numbers,
- $\mathbb{R}$: the set of real numbers,
- $\mathbb{C}$: the set of complex numbers,
- $\mathbb{Z}^+ = \{0, 1, 2, \dots\}$,
- $\mathbb{Q}^+ = \{r \in \mathbb{Q} | r > 0\}$,
- $\mathbb{R}^+ = \{r \in \mathbb{R} | r > 0\}$,
- $\mathbb{Z}^* = \mathbb{Z} \setminus \{0\}$,
- $\mathbb{Q}^* = \mathbb{Q} \setminus \{0\}$,
- $\mathbb{R}^* = \mathbb{R} \setminus \{0\}$,
- $\mathbb{C}^* = \mathbb{C} \setminus \{0\}$,
- $M_{m \times n}(F)$: the set of all $m \times n$ matrices with entries in a field F,
- $M_n(F) = M_{n \times n}(F)$,
- I_n: the identity matrix of size n.

1. BASIC THEORY ON RINGS

The conceptualization of rings spanned the 1870s to the 1920s, with key contributions by Richard Dedekind (1831–1916), David Hilbert (1862–1943), Abraham Fraenkel (1891–1965), and Emmy Noether (1882–1935). Rings were first formalized as a generalization of Dedekind domains that occur in number theory, and of polynomial rings and rings of invariants that occur in algebraic geometry and invariant theory. They later proved useful in other branches of mathematics such as geometry and analysis.

In this chapter we will first recall some definitions and results about rings and fields from an elementary course on abstract algebra, including basic properties of rings and isomorphism theorems. Proofs of results from an elementary course on abstract algebra will often be omitted in class although we contain details in this book. Then we introduce some basic tools for this book, for example, the field of quotients of an integral domain, basic properties for irreducible polynomials over a field.

1.1. Basic properties of rings.

In this section we will recall some definitions and results from Abstract Algebra. Most contents of the first two sections and Section 1.5 can be found in [LZ].

Definition 1.1.1. A **group** G *is a set with a binary operation (generally called product)* $G \times G \to G$, $(a, b) \mapsto a \cdot b$ *for* $a, b \in G$ *satisfying the following three axioms.*

(G1) The product is associative:

$$a \cdot (b \cdot c) = (a \cdot b) \cdot c, \forall a, b, c \in G.$$

(G2) There is an **identity element** e *in* G:

$$a \cdot e = e \cdot a = a, \forall a \in G.$$

(G3) Each element a *in* G *has an* **inverse** $a^{-1} \in G$:

$$a \cdot a^{-1} = a^{-1} \cdot a = e.$$

1

We generally denote the above group G as $(G, \cdot, e)$, and $a \cdot b$ is denoted by ab for convenience.

Definition 1.1.2. *A group G is called* **abelian** *if $ab = ba$ for all $a, b \in G$.*

Example 1.1.1. *We have the following familiar examples of groups:* $(\mathbb{Z}, +, 0)$, $(\mathbb{Q}, +, 0)$, $(\mathbb{R}, +, 0)$, $(\mathbb{C}, +, 0)$, $(\mathbb{Q}^*, \cdot, 1)$, $(\mathbb{R}^*, \cdot, 1)$, $(\mathbb{C}^*, \cdot, 1)$.

Example 1.1.2. *Let X be a nonempty set. Denote S_X by the set of all bijections from X to itself. So any $f \in S_X$ has an inverse map in S_X denoted by f^{-1}. Clearly the identity map id_X belongs to S_X. For any $f, g \in S_X$, define the composition $g \cdot f$ by $(g \cdot f)(x) = g(f(x))$, $x \in X$. This composition gives a binary operation*

$$S_X \times S_X \to S_X, \ (g, f) \mapsto g \cdot f,$$

satisfying

$$(f \cdot g) \cdot h = f \cdot (g \cdot h),$$
$$\mathrm{id}_X \cdot f = f \cdot \mathrm{id}_X = f,$$
$$f^{-1} \cdot f = f \cdot f^{-1} = \mathrm{id}_X,$$

for all $f, g, h \in S_X$. Thus $(S_X, \cdot, \mathrm{id}_X)$ is a group, called the **symmetric group** *on X. We simply write S_X as S_n if $X = \{1, 2, \cdots, n\}$.*

Definition 1.1.3. *A subset $H \subseteq G$ is a* **subgroup** *of a group G if H is a group with the same operation as G, i.e.:*

(1). H is closed under "$\cdot$",
(2). $e \in H$,
(3). $a^{-1} \in H$, for all $a \in H$.

We denote this by writing $H \leq G$ (or $G \geq H$). The trivial subgroups of G are $\{e\}, G \leq (G, \cdot, e)$. The meaning for $H < G$ (or $G > H$) is clear.

Definition 1.1.4. *Let $(G, \cdot, e)$ be a group, $H \leq G$ and $a \in G$. The* **(left) coset** *of H containing a is the subset $aH = \{ah : h \in H\}$ of G. We denote*

$$\frac{G}{H} = \{aH : a \in G\},$$

which is also denoted as G/H.

Definition 1.1.5. *A subgroup N of a group G is called a* **normal subgroup** *of G if $g^{-1}xg \in N$ for all $x \in N$ and all $g \in G$. We write*

$N \trianglelefteq G$ to indicate that N is a normal subgroup of G. By $N \triangleleft G$ we have the obvious meaning.

Theorem 1.1.6. *Let G be a group and N be a normal subgroup of G. Define a binary operation on G/N by*

$$xN \cdot yN = xyN$$

for $x, y \in G$. With this multiplication, G/N is a group.

Definition 1.1.7. *If G is a group and N is a normal subgroup of G, we call group G/N (with the above multiplication) the* **quotient group of G by N.**

Definition 1.1.8. *A* **ring** *R is a set with two binary operations, $+$ and $\cdot$, satisfying:*
(R1) $(R, +, 0)$ is an abelian group,
(R2) associativity under multiplication: $(ab)c = a(bc)$ for all a, b, $c \in R$,
(R3) distributivities: $a(b + c) = ab + ac$ and $(a + b)c = ac + bc$ for all $a, b, c \in R$.

The ring R is usually denoted as $(R, +, \cdot)$. From now on in this book we always assume that R is a ring. A ring R is called **commutative** if $ab = ba$ for all $a, b \in R$.

Proposition 1.1.9. *For any $r, s \in R$, we have*
(1). $r0 = 0r = 0$,
(2). $(-r)s = r(-s) = -(rs)$.

Proof. (1). $r0 = r(0 + 0) = r0 + r0$. Adding $-(r0)$ to both sides, we get:

$$0 = r0 - (r0) = r0 + r0 - r0 = r0.$$

Similarly, $0r = 0$.
(2). $0 = 0s$ by (1) and

$$0 = 0s = (-r + r)s = (-r)s + rs.$$

Add $-(rs)$ to both sides to get $-(rs) = (-r)s$. Similarly, $r(-s) = -(rs)$. $\qquad \square$

Example 1.1.3. *(a). We can easily see that the following number systems are rings:*

$$(\mathbb{Z}, +, \cdot), (\mathbb{Q}, +, \cdot), (\mathbb{C}, +, \cdot), (\mathbb{R}, +, \cdot), (\mathbb{Z}_n, +, \cdot).$$

(b). The set $(M_n(\mathbb{R}), +, \cdot)$ with matrix multiplication and addition is a ring. Note that if we replace $\mathbb{R}$ with any number system this still holds true. For example, $(M_n(\mathbb{Z}), +, \cdot)$ is a ring.

Example 1.1.4.   *(a). Fix $m \in \mathbb{N}$. For any $n \in \mathbb{Z}$, write $\bar{n} = \{n + mk : k \in \mathbb{Z}\}$. Define*

$$\bar{n}_1 + \bar{n}_2 = \overline{n_1 + n_2}, \ and \ \bar{n}_1 \cdot \bar{n}_2 = \overline{n_1 n_2}.$$

*The classes $\bar{0}, \bar{1}, \cdots, \overline{m-1}$ are called **residues modulo m**. The set $\{\bar{0}, \bar{1}, \cdots, \overline{m-1}\}$ is denoted by $\mathbb{Z}_m$ or by $\mathbb{Z}/m\mathbb{Z}$. Then $(\mathbb{Z}_m, +, \cdot)$ is a commutative ring.*

(b). The set of polynomials in x with coefficients in $\mathbb{Q}$ (or in $\mathbb{R}$ or $\mathbb{C}$)

$$\mathbb{Q}[x] = \{f(x) = a_0 + a_1 x + \cdots + a_n x^n : n \in \mathbb{N}, a_i \in \mathbb{Q}\}$$

*with usual addition and multiplication is a commutative ring. If $a_n \neq 0$ then n is the **degree** of $f(x)$, denoted by $\deg(f(x)) = n$ and we define $\deg(0) = -\infty$.*

Definition 1.1.10. *A **subring** of a ring R is a subset S of R which is a ring under the same addition and multiplication as in R, denoted by $S \leq R$.*

The meaning for $S < R$ (or $R > S$) is clear.

Proposition 1.1.11. *A non-empty subset S of a ring R is a subring of R if and only if $a + b, ab, -a \in S$ for any $a, b \in S$.*

Proof. ($\Rightarrow$). Clearly, a subring has these properties.

($\Leftarrow$). If S is a non-empty subset of R such that $a + b, ab, -a \in S$ for any $a, b \in S$, then $(S, +)$ is a subgroup of $(R, +)$ (from group theory), and S is closed under multiplication. Associativity (R2) and distributivities (R3) hold for S because they hold for R. $\square$

Definition 1.1.12. *Let d be an integer which is not a square. Define*

$$\mathbb{Z}[\sqrt{d}] = \{a + b\sqrt{d} : a, b \in \mathbb{Z}\}.$$

*Call $\mathbb{Z}[\sqrt{-1}] = \{a + b\sqrt{-1}, a, b \in \mathbb{Z}\}$ the ring of **Gaussian integers**.*

Proposition 1.1.13. *Let d be an integer which is not a square. Then $(\mathbb{Z}[\sqrt{d}], \cdot, +)$ is a ring. Moreover, $m + n\sqrt{d} = m' + n'\sqrt{d}$ in $\mathbb{Z}[\sqrt{d}]$ if and only if $m = m'$ and $n = n'$.*

Proof. We will show that $(\mathbb{Z}[\sqrt{d}], \cdot, +) < (\mathbb{C}, \cdot, +)$. Consider $m, n, a, b \in \mathbb{Z}$. Then we have:

Closure under addition: $(m + n\sqrt{d}) + (a + b\sqrt{d}) = (m + a) + (n + b)\sqrt{d}$.

Closure under multiplication: $(m + n\sqrt{d})(a + b\sqrt{d}) = ma + nbd + (mb + na)\sqrt{d}$.

Also, $-(m + n\sqrt{d}) = (-m) + (-n)\sqrt{d}$.

Hence $(\mathbb{Z}[\sqrt{d}], \cdot, +) < (\mathbb{C}, \cdot, +)$.

Finally, if $m + n\sqrt{d} = m' + n'\sqrt{d}$, then if $n \neq n'$ we write $\sqrt{d} = \frac{m-m'}{n'-n}$ which is not possible since $\sqrt{d}$ is not a rational number. Therefore, $n = n'$ hence $m = m'$. □

Definition 1.1.14. *An nonzero element a of a ring R is called a* **left zero divisor** *if there exists nonzero $b \in R$ such that $ab = 0$. We can similarly define a* **right zero divisor** *of a ring. The multiplicative identity element of a ring R, if it exists, is denoted by 1 and is called the* **unity***. A* **unital ring** *R is ring with unity 1 such that $1 \neq 0$.*

If $0 = 1$, then $x \cdot 1 = x$ and so $x = x \cdot 1 = x \cdot 0 = 0$. Hence if $0 = 1$ then $R = \{0\}$. Note that, in some other books, the rings we defined are called rng (or pseudo-ring), and unital rings are called rings.

Definition 1.1.15. *If there exists a positive integer n such that $na = 0$ for all $a \in R$, then the least such positive integer is called the* **characteristic** *of R, denoted by $\mathrm{char}(R) = n$. If no such positive integer exists, then the characteristic of R is 0, denoted by $\mathrm{char}(R) = 0$.*

Example 1.1.5. *We know that $(\mathbb{C}, +, \cdot)$ is a ring with characteristic $\mathrm{char}(\mathbb{C}) = 0$, and $(\mathbb{Z}_n, +, 0)$ has characteristic $\mathrm{char}(\mathbb{Z}_n) = n$.*

Example 1.1.6. *Let R be the subring of $M_2(\mathbb{Z}_3)$ consisting of matrices of the form*

$$\begin{bmatrix} a & b \\ 0 & 0 \end{bmatrix}, a, b \in \mathbb{Z}_3.$$

We can see that $\begin{bmatrix} 1 & 0 \\ 0 & 0 \end{bmatrix}$ is a right zero divisor but not a left zero divisor.

Definition 1.1.16. *A ring R is called an* **integral domain** *if*
(1) R is commutative,
(2) R is unital, and
(3) R has no zero divisors.

For example $(\mathbb{Z}, +, \cdot)$, $(\mathbb{Q}, +, \cdot)$, $(\mathbb{R}, +, \cdot)$, $(\mathbb{C}, +, \cdot)$, $(\mathbb{Z}[\sqrt{d}] +, \cdot)$, $(\mathbb{Q}[x] +, \cdot)$ are integral domains.

If R is an integral domain (or any commutative ring), then $R[x]$ denotes the set of polynomials in x with coefficients from R with usual addition and multiplication. Clearly $R[x]$ is a commutative ring.

Proposition 1.1.17. *If R is an integral domain, then $R[x]$ is also.*

Proof. We need only to check that there are no zero divisors. For contradiction, assume that

$$f(x) = a_0 + a_1 x + \cdots + a_m x^m, g(x) = b_0 + \cdots + b_n x^n$$

are elements of $R[x]$ such that $f(x)g(x)$ is the zero polynomial. Without loss of generality assume that $a_m \neq 0, b_n \neq 0$, i.e., m $=$ deg$f(x)$, $n =$ deg$g(x)$. Then $f(x)g(x) = a_0 b_0 + \cdots + a_m b_n x^{m+n}$.

Since R is an integral domain, $a_m b_n \neq 0$. Therefore we get a contradiction, hence $f(x)g(x) \neq 0$. $\qquad\square$

From the above proof we see that for any $f(x), g(x) \in R[x]$,

$$\deg(fg) = \deg(f) + \deg(g),$$

if R is an integral domain.

Definition 1.1.18. *Let R be a unital ring. An element $a \in R$ is a* **unit** *if there exists $a^{-1} \in R$ such that $aa^{-1} = a^{-1}a = 1$. The element a^{-1} is unique if it exists, and is called the* **inverse** *of a. The* **unit group** *of R is the set consisting of all units of R is denoted as $\mathcal{U}(R)$.*

Definition 1.1.19. *Let R be a ring and let $a, b \in R$. We say that then a is a* **factor** *of b or a* **divides** *b if there exists $c \in R$ such that $b = ac$, denoted by $a|b$. We read $a \nmid b$ as "a does not divide b". Notice that $a|0$ for any element $a \in R$.*

Definition 1.1.20. *Let R be a unital ring. A non-unit element $a \in R$ is* **irreducible** *if $a = bc$ for some $b, c \in R$ implies that b or c is a unit.*

Note that irreducible polynomials in $\mathbb{Q}[x]$ are exactly irreducibles in the polynomial ring $\mathbb{Q}[x]$.

Theorem 1.1.21. *In $(\mathbb{Z}_n, +, \cdot)$, the 0-divisors are the nonzero elements that are not coprime to n.*

Proof. Let $m \in \mathbb{Z}_n \setminus \{0\}, 1 \leq m \leq n - 1$.

Case 1: Suppose that $\gcd(m, n) = d \neq 1$ (i.e., m and n are not relatively prime). Then $\dfrac{m}{d}, \dfrac{n}{d} \in \mathbb{Z}$. Then it follows that $0 < \dfrac{m}{d}, \dfrac{n}{d} < n$, so $\dfrac{m}{d}, \dfrac{n}{d} \neq 0$ in $\mathbb{Z}_n$. Then $m \cdot \dfrac{n}{d} = \dfrac{m}{d} \cdot n = 0$ in $\mathbb{Z}_n$. Thus m is a 0-divisor.

Case 2: Now suppose $\gcd(m, n) = 1$ (i.e., m and n are relatively prime). Let $s \in \mathbb{Z}_n$ so that $ms = 0$ in $\mathbb{Z}_n$, i.e., $n | ms$. So $n | s$. So $s = 0$ in $\mathbb{Z}_n$. So m is not a 0-divisor. $\qquad \square$

Proposition 1.1.22. *Let n be a positive integer. Then $(\mathbb{Z}_n, +, \cdot)$ is an integral domain if and only if n is prime.*

Proof. This follows from the previous theorem. $\qquad \square$

Proposition 1.1.23. *Every integral domain R satisfies the cancellation property: if $ax = ay$ and $a \neq 0$ then $x = y$ for all $x, y, a \in R$.*

Proof. If $ax = ay$ then $a(x - y) = 0$. Since R has no zero divisors and $a \neq 0$, we conclude that $x - y = 0$, so that $x = y$. $\qquad \square$

Definition 1.1.24. *A unital ring D is called a **division ring** if $\mathcal{U}(D) = D \setminus \{0\}$. A unital ring F is a **field** if it is a commutative division ring.*

A first step towards the notion of a field was made in 1770 by Joseph-Louis Lagrange (1736–1813). The first clear definition of an abstract field (1893) is due to Heinrich Martin Weber (1842–1913).

Example 1.1.7. *(1). We have the following familiar examples of fields: $\mathbb{Q}, \mathbb{R}, \mathbb{C}, \mathbb{Z}_2, \mathbb{Z}_3$, where $\mathbb{Z}_2 = \{\overline{0}, \overline{1}\}, \mathbb{Z}_3 = \{0, 1, 2\} = \{0, 1, -1\}$.*

(2). For a non-square $d \in \mathbb{Q}$, then $\mathbb{Q}[\sqrt{d}] = \{x + y\sqrt{d} : x, y \in \mathbb{Q}\}$ is a field.

It is easy to see that $\mathbb{Q}[\sqrt{d}]$ is a subring of $\mathbb{C}$. Assuming $x \neq 0$ or $y \neq 0$, we compute

$$\frac{1}{x + y\sqrt{d}} = \frac{x - y\sqrt{d}}{(x - y\sqrt{d})(x + y\sqrt{d})} = \frac{x - y\sqrt{d}}{x^2 - y^2 d} \in \mathbb{Q}[\sqrt{d}].$$

Note that $x^2 - y^2 d \neq 0$ since d is not a square of a rational number.

Definition 1.1.25. *A subset S of a field F is a* **subfield** *if S is a field with the same addition and multiplication, denoted by $S \leq F$. The obvious meaning of $S < F$ is clear.*

To check that a subset S of a field F is a subfield, it is enough to check that $0, 1, a - b, ab^{-1} \in S$ (if $b \neq 0$) for any $a, b \in S$.

Definition 1.1.26. *Let F be a subfield of a field K and $\alpha_1, \cdots, \alpha_n \in K$. The smallest subfield of K containing F and $\alpha_1, \cdots, \alpha_n$ is denoted by $F(\alpha_1, \alpha_2, \cdots, \alpha_n)$.*

Example 1.1.8. *This notation agrees with $\mathbb{Q}(\sqrt{d}) = \{a + b\sqrt{d} : a, b \in \mathbb{Q}\}$ where d is a non-square rational number. Let's check that $\mathbb{Q}(\sqrt{d})$ is indeed the smallest subfield of $\mathbb{C}$ containing $\mathbb{Q}$ and $\sqrt{d}$. The smallest subfield must contain all numbers like $a\sqrt{d}, a \in \mathbb{Q}$, since it is closed under $\cdot$, and hence also all numbers like $a + a'\sqrt{d}, a, a' \in \mathbb{Q}$, since closed under $+$. So $\{a + b\sqrt{d} : a, b \in \mathbb{Q}\} \subset \mathbb{Q}(\sqrt{d})$. We also know that $\{a + b\sqrt{d} : a, b \in \mathbb{Q}\}$ is a field.*

Similarly we can consider $\mathbb{Q}(\sqrt{d_1}, \sqrt{d_2})$, and more complicated fields.

Theorem 1.1.27. *Every field is an ID.*

Proof. Let F be a field. Then $0 \neq 1 \in F$ and F is commutative. Suppose on the contrary that $a \in F$ is a 0-divisor. Then $a \neq 0$ and there exists $b \neq 0$ such that $ab = 0$. Since F is a field, then a^{-1} exists and so
$$0 \neq b = a^{-1}(ab) = a^{-1}0 = 0$$
and is clearly a contradiction. So F cannot have any 0-divisors. So F must be an ID. $\qquad\square$

The following theorem was proved by Joseph Wedderburn (1882–1948) in 1905.

Theorem 1.1.28 (Wedderburn's Little Theorem). *Every finite ID is a field.*

Proof. Let $D = \{0, 1, a_1, \cdots, a_n\}$ be a finite ID. We need to show that the unit group $\mathcal{U}(D) = D \backslash \{0\}$. Let $a \in D \backslash \{0\}$. Since $ax = ay$ implied $x = y$ we have
$$\underbrace{\{a1, aa_1, \cdots, aa_n\}}_{n+1 \text{ elements}} = \{1, a_1, \cdots, a_n\}.$$

Since these two sets are equal, there exists $b \in D \setminus \{0\}$ such that $ab = 1$. So $b = a^{-1}$ and so D must be a field. $\qquad\square$

Corollary 1.1.29. *The ring $\mathbb{Z}_m$ is a field if and only if m is prime.*

Proof. If m is not prime then we know that $\mathbb{Z}_m$ has zero divisors, hence is not a field.

If m is a prime, then $\mathbb{Z}_m$ is a finite integral domain, hence a field by the previous theorem. $\qquad\square$

1.2. Isomorphism theorems.

Definition 1.2.1. *A subring I of a ring R is called an **ideal** if $ar, ra \in I$ for all $a \in I, r \in R$. If I is an ideal of R we denote this fact by $I \trianglelefteq R$. By $I \triangleleft R$ (or $I \subsetneqq R$) we mean $I \trianglelefteq R$ and $I \neq R$.*

Proposition 1.2.2. *A non-empty subset I of a ring R is an ideal of R if and only if $a - b, ar, ra \in I$ whenever $a, b \in I$ and $r \in R$.*

Proof. This is easy to see. $\qquad\square$

Definition 1.2.3. *Let I be an ideal of a ring R and $x \in R$. The **coset** of I in R containing x is*

$$x + I = \{x + i : i \in I\}.$$

When dealing with cosets, it is more important to realize that, in general, a given coset can be represented in more than one way. The next lemma shows how the coset representatives are related.

Lemma 1.2.4. *Let R be a ring with an ideal I and $x, y \in R$. Then $x + I = y + I$ if and only if $x - y \in I$.*

Proof. We omit the detailed proof since it is easy. $\qquad\square$

Theorem 1.2.5. *Let $I \trianglelefteq R$. Then $R/I = \{a + I : a \in R\}$ is a ring with*

$$(a + I) + (b + I) = (a + b) + I, (a + I)(b + I) = ab + I, \forall a, b \in R.$$

*The ring $(R/I, +, \cdot)$ is call the **quotient ring** of R by I.*

Proof. The proof is fairly standard and can be found in any Abstract Algebra book like [LZ]. $\qquad\square$

The zero element of R/I is $I = 0 + I = a + I$ for any $a \in I$. If S is a subset of R with $S \supseteq I$ we denote by S/I the subset $\{s + I : s \in S\}$ of R/I.

Definition 1.2.6. *A map θ of a ring R into a ring S is said to be a (ring)* **homomorphism** *if $\theta(x + y) = \theta(x) + \theta(y)$ and $\theta(xy) = \theta(x)\theta(y)$ for all $x, y \in R$.*

The map $\theta : R \to S$ defined by $\theta(r) = 0$ for all $r \in R$ is a homomorphism. It is called the **zero homomorphism**.

The map $\phi : R \to R$ defined by $\phi(r) = r$ for all $r \in R$ is also a homomorphism. It is called the **identity homomorphism**.

Let $I \trianglelefteq R$. Then $\sigma : R \to R/I$ defined by $\sigma(x) = x + I$ for all $x \in R$ is a homomorphism of R onto R/I. This is called the **natural (or canonical) homomorphism**.

Proposition 1.2.7. *Let R, S be rings and $\theta : R \to S$ a homomorphism. Then*

(1). $\theta(0_R) = 0_S$,
(2). $\theta(-r) = -\theta(r)$ for all $r \in R$,
(3). $K = \{x \in R : \theta(x) = 0_S\}$ is an ideal of R,
(4). $\theta R = \{\theta(r) : r \in R\}$ is subring of S,
(5). $\theta^{-1}(\theta(x)) = x + K$ for any $x \in R$.

Proof. The proof is standard. □

In the above theorem, K is called the **kernel** of θ and θR is called the (homomorphic) **image** of R. The ideal K is sometimes denoted by $\ker(\theta)$.

Definition 1.2.8. *Let θ be a homomorphism of a ring R into a ring S. Then θ is called an* **isomorphism** *if θ is a one to one and onto map. We say that R and S are isomorphic rings if there is an isomorphism $\theta : R \to S$, denote this by $R \cong S$.*

Lemma 1.2.9. *Let $N \trianglelefteq R$ and let*

$$[\![R, N]\!] = \{K \leq R \,|\, K \supseteq N\}, \quad [\![R/N]\!] = \{K' \leq R/N\}.$$

(1). $\Psi : [\![R, N]\!] \to [\![R/N]\!]$ defined by $K \mapsto K/N$ is a bijection.
(2). For $K \in [\![R, N]\!]$ we have that $K \trianglelefteq R$ if and only if $K/N \trianglelefteq R/N$.

Proof. (1). Consider the homomorphism $\gamma : R \to R/N, x \mapsto x + N$, which is onto. For any $K \in [\![R, N]\!]$, we see that $\Psi(K) = \gamma(K) = K/N$.

For $K' \in [\![R/N]\!]$, let $K = \gamma^{-1}(K')$. Then $N \leq K \leq R$. We see that $\Psi(K) = \gamma(\gamma^{-1}(K')) = K'$ since γ is onto. So Ψ is onto.

Let $K_1, K_2 \in [\![R, N]\!]$ such that $\Psi(K_1) = \Psi(K_2)$. Then $\gamma(K_1) = \gamma(K_2)$, and

$$K_1 = K_1 + N = \gamma^{-1}(\gamma(K_1)) = \gamma^{-1}(\gamma(K_2)) = K_2 + N = K_2.$$

So Ψ is 1-1. Thus $\Psi : [\![R, N]\!] \to [\![R/N]\!]$ is a bijection.

(2). We see that $K \trianglelefteq R$ if and only if $rK, Kr \subseteq K$ for any $r \in R$, if and only if $r(K/N) = (r + N)(K/N), (K/N)r = (K/N)(r + N) \subseteq K/N$ for any $r \in R$, if and only if $K/N \trianglelefteq R/N$. $\qquad \square$

Theorem 1.2.10 (First Isomorphism Theorem). *Let $\theta : R \to S$ be a ring homomorphism. Then $\theta R \cong R/I$ where $I = \ker \theta$.*

Proof. Defined $\sigma : R/I \to \theta R$ by $\sigma(x + I) = \theta(x)$ for all $x \in R$. The map σ is well defined since for $x, y \in R$,

$$x+I = y+I \Longleftrightarrow x-y \in I = \ker \theta \Longleftrightarrow \theta(x-y) = 0 \Longleftrightarrow \theta(x) = \theta(y).$$

Clearly σ is 1-1 and onto. Then σ is the required isomorphism. $\qquad \square$

Theorem 1.2.11 (Second Isomorphism Theorem). *Let I be an ideal and L a subring of a ring R. Then $L \cap I \trianglelefteq L$, $I \trianglelefteq L + I$, and $L/(L \cap I) \cong (L + I)/I$.*

Proof. It is easy to see that $L \cap I \trianglelefteq L$, $I \trianglelefteq L + I$.

Let σ be the natural homomorphism $R \to R/I$. Restrict σ to the ring L. We have $\sigma L = (L + I)/I$. The kernel of σ restricted to L is $L \cap I$. Now apply previous theorem. $\qquad \square$

Theorem 1.2.12 (Third Isomorphism Theorem). *Let $I, K \trianglelefteq R$ such that $I \subseteq K$. Then*

$$(R/I)/(K/I) \cong R/K.$$

Proof. Since $K/I \triangleleft R/I$, so $(R/I)/(K/I)$ is defined. Define a map

$$\gamma : R/I \to R/K, \quad \gamma(x + I) = x + K, \forall x \in R.$$

The map γ is easily seen to be well defined and a homomorphism onto R/K. Further,

$$\begin{aligned}
\gamma(x + I) = K &\Longleftrightarrow \quad x + K = K \\
&\Longleftrightarrow \quad x \in K \\
&\Longleftrightarrow \quad x + I \in K/I.
\end{aligned}$$

Therefore $\ker \gamma = K/I$. Now apply the first isomorphism theorem. $\qquad \square$

The isomorphism theorems were formulated in some generality for homomorphisms of modules by Emmy Noether in her paper "Abstrakter Aufbau der Idealtheorie in algebraischen Zahlund Funktionenkrpern", which was published in 1927 in "Mathematische Annalen". (See Theorems 3.1.10–3.1.12.) Three years later, B.L. van der Waerden (1903–1996) published his influential "Algebra", the first abstract algebra textbook that took the groups-rings-fields approach to the subject. The three isomorphism theorems, called homomorphism theorem, and two laws of isomorphism when applied to rings, appeared explicitly.

Definition 1.2.13. *Let $R_1, \ldots, R_n$ be rings. We define the* **external direct sum** *S to be the set of all n-tuples $\{(r_1, \ldots, r_n) : r_i \in R_i\}$. On S we define addition and multiplication component wise. This makes S a ring. We write $S = R_1 \oplus \cdots \oplus R_n$.*

The set $(0, \ldots, 0, R_j, 0, \ldots, 0)$ is an ideal of S. Clearly S is the internal direct sum of these ideals. But $(0, \ldots, R_j, \ldots 0) \cong R_j$. Because of this S can be considered as a ring in which the R_j are ideals and S is their internal direct sum. Also in internal direct sum we can consider $I_1 \oplus \cdots \oplus I_n$ to be the external direct sum of the rings I_j. Hence, in practice, we do not need to distinguish between external and internal direct sums.

Definition 1.2.14. *Let $\{I_\lambda\}_{\lambda \in \Lambda}$ be a collection of ideals of a ring R. We define their* **sum** *to be*

$$\sum_{\lambda \in \Lambda} I_\lambda = \{x_1 + \cdots + x_k : x_i \in I_{\lambda_i}, \lambda_k \in \Lambda\}.$$

That is, the sum is the collection of finite sums of elements of the I_λ's.

We say that the sum of the I_λ's is **direct** *if each element of $\sum_{\lambda \in \Lambda} I_\lambda$ is uniquely expressible as*

$$x_1 + \cdots + x_k, \text{ with } x_i \in I_{\lambda_i}.$$

In this case we denote this sum as $\oplus_{\lambda \in \Lambda} I_\lambda$ or $I_1 \oplus \cdots \oplus I_n$ if Λ is finite.

Proposition 1.2.15. *The sum $\sum_{\lambda \in \Lambda} I_\lambda$ is direct if and only if*

$$I_\mu \cap \left(\sum_{\lambda \in \Lambda, \lambda \neq \mu} I_\lambda \right) = 0, \text{ for all } \mu \in \Lambda.$$

Proof. The proof is standard and we omit the details.   □

Example 1.2.1. *If ϕ is a nonzero ring homomorphism from the real numbers $\mathbb{R}$ to $\mathbb{R}$, show that ϕ is the identity map. (Hint: show $x > 0$ implies $\phi(x) > 0$.)*

Solution. Since $\phi \neq 0$ we see that ϕ is one-to-one, $\phi(\pm 1) = \pm 1$ and $\phi(0) = 0$.

Let $x > 0$ be a positive real number. Then there exists $y \in \mathbb{R}$ such that $x = y^2$. Hence $\phi(x) = \phi(y)^2 > 0$.

If $a < b$, then $b - a > 0$. Hence $\phi(b) - \phi(a) = \phi(b - a) > 0$ by the above. Hence $\phi(a) < \phi(b)$. This means that for $a, b \in \mathbb{R}$,

$$a < b \implies \phi(a) < \phi(b).$$

If $n \in \mathbb{N}$, using $n = 1 + \cdots + 1$, we have

$$\phi(n) = \phi(1 + \cdots + 1) = \phi(1) + \cdots + \phi(1) = 1 + \cdots + 1 = n, \forall n \in \mathbb{N},$$

and $\phi(-n) = -n$ for all $n \in \mathbb{N}$. So

$$\phi(n) = n, \forall n \in \mathbb{Z}.$$

Now, any rational number is of the form $r = ac^{-1}$ for $a, c \in \mathbb{Z}$ with $c \neq 0$. It follows that

$$\phi(r) = \phi(ac^{-1}) = \phi(a)\phi(c)^{-1} = ac^{-1} = r, \forall r \in \mathbb{Q}.$$

Let $a \in \mathbb{R}$ and suppose $\phi(a) \neq a$, say $\phi(a) > a$. We know that there is $q \in \mathbb{Q}$ such that $a < q < \phi(a)$. Hence, $\phi(a) < \phi(q) = q < \phi(a)$. This is a contradiction. Thus $\phi(a) = a$ for all $a \in \mathbb{R}$.   □

1.3. The field of quotients of an integral domain.

If every nonzero element in an integral domain D has a multiplicative inverse, then D is a field. It is the purpose of this section to show that every integral domain can be regarded as subring of a field, a field of quotients of the integral domain. This field will be a minimal field containing the integral domain. For example, the integers are contained in the field $\mathbb{Q}$, whose elements can all be expressed as quotients of integers.

We can follow the steps by the way $\mathbb{Q}$ can be formed from $\mathbb{Z}$.

Let D be an integral domain that we desire to enlarge to a field of quotients F. We take four steps to obtain F as follows.

1. Define the elements of F.
2. Define addition and multiplication on F.

3. Show that F is a field under these operations.

4. Show that D can be considered a subring of F.

Step 1. Consider $S = \{(a, b) | a, b \in D, b \neq 0\}$.

Definition 1.3.1. *Two elements* $(a, b), (c, d) \in S$ *are* **equivalent**, *denoted by* $(a, b) \sim (c, d)$, *if* $ad = bc$.

Theorem 1.3.2. *The relation* $\sim$ *on* S *is an equivalence relation.*

Proof. Reflexive: $(a, b) \sim (a, b)$ since $ab = ba$ (D is an integral domain).

Symmetric: $(a, b) \Longleftrightarrow ad = bc \Longleftrightarrow (c, d) \sim (a, b)$.

Transitive: If $(a, b) \sim (c, d)$ and $(c, d) \sim (r, s)$, then $ad = bc$ and $cs = dr$. We have

$$asd = sad = sbc = bcs = bdr = brd.$$

Now $d \neq 0$, and D is an integral domain, so cancellation is valid. Hence from $asd = brd$ we obtain $as = br$, so that $(a, b) \sim (r, s)$. $\quad\square$

We now know that $\sim$ gives a partition of S into equivalence classes. We shall let $\frac{a}{b}$ be the equivalence class of (a, b) in S under the relation $\sim$, i.e.,

$$\frac{a}{b} = \{(c, d) \in S : (c, d) \sim (a, b)\}.$$

Let

$$F = \left\{ \frac{a}{b} : (a, b) \in S \right\}.$$

Step 2. Define addition and multiplication in F. Observe that if $D = \mathbb{Z}$ and $\frac{a}{b}$ is viewed as $a/b \in \mathbb{Q}$, these definitions applied to $\mathbb{Q}$ give the usual operations.

Theorem 1.3.3. *For* $\frac{a}{b}, \frac{c}{d} \in F$, *the operations*

$$\frac{a}{b} \cdot \frac{c}{d} = \frac{ac}{bd}, \quad \frac{a}{b} + \frac{c}{d} = \frac{ad + bc}{bd}$$

are well-defined on F.

Proof. Since $\frac{a}{b}, \frac{c}{d} \in F$, then (a, b) and (c, d) are in S, and $bd \neq 0$. So

$$(ad + bc, bd), (ac, bd) \in S.$$

Thus $\frac{ac}{bd}, \frac{ad+bc}{bd} \in F$.

To see these operations of addition and multiplication are well defined, suppose that $\frac{a_1}{b_1} = \frac{a}{b}$ and $\frac{c_1}{d_1} = \frac{c}{d}$. Then

$$a_1 b = b_1 a, \quad c_1 d = d_1 c. \tag{1.1}$$

We must show that

$$\frac{a_1 d_1 + b_1 c_1}{b_1 d_1} = \frac{ad + bc}{bd}, \text{ and } \frac{a_1 c_1}{b_1 d_1} = \frac{ac}{bd},$$

i.e.,

$$(a_1 d_1 + b_1 c_1)bd = b_1 d_1 (ad + bc), \text{ and } a_1 c_1 bd = b_1 d_1 ac.$$

These can be verified by using (1.1). Now we complete the proof. $\square$
From this theorem we see that,

$$\frac{ab}{ac} = \frac{b}{c}, \forall a, b, c \in D \text{ with } ac \neq 0.$$

Step 3. Check that F is a field under these operations.

Theorem 1.3.4. *The above defined $(F, +, \cdot)$ from D is a field.*

Proof.

(1). Addition in F is commutative: Since $\frac{a}{b} + \frac{c}{d} = \frac{bc+da}{bd}$ and $\frac{c}{d} + \frac{a}{b} = \frac{ad+bc}{bd}$. So $\frac{a}{b} + \frac{c}{d} = \frac{c}{d} + \frac{a}{b}$.

(2). Addition is associative. This is easy to verify.

(3). The element $\frac{0}{1}$ is an identity element for addition in F. This is clear.

(4). The element $\frac{-a}{b}$ is an additive inverse for $\frac{a}{b}$ in F. This is clear.

(5). Multiplication in F is associative. This is easy to verify.

(6). Multiplication in F is commutative. This is easy to verify.

(7). The distributive laws hold in F. This is easy to verify.

(8). The element $\frac{1}{1}$ is a multiplicative identity element in F. This is clear.

(9). If $\frac{a}{b} \in F$ is not the additive identity element, then $a \neq 0$ in D and $\frac{b}{a}$ is a multiplicative inverse for $\frac{a}{b}$:

Let $\frac{a}{b} \in F$. If $a = 0$, then $\frac{a}{b} = \frac{b0}{b1} = \frac{0}{1}$. But $\frac{0}{1}$ is the additive identity by (3). Thus if $\frac{a}{b} \neq \frac{0}{1}$ in F, we have $a \neq 0$. Now $\frac{a}{b}\frac{b}{a} = \frac{ab}{ba} = \frac{1}{1}$. Thus

$$\frac{a}{b}\frac{b}{a} = \frac{1}{1},$$

and $\frac{1}{1}$ is the multiplicative identity by (8). So $(F, \cdot, +)$ is a field. $\square$
This completes Step 3.

Step 4. Show that F can be regarded as containing D.

Theorem 1.3.5. *The map* $\iota : D \to F$ *given by* $\iota(a) = \frac{a}{1}$ *is an isomorphism of* D *with a subring of* F.

Proof. For $a, b \in D$, we have

$$\iota(a) + \iota(b) = \frac{a}{1} + \frac{b}{1} = \frac{a1 + 1b}{1} = \frac{a + b}{1} = \iota(a + b),$$

$$\iota(a)\iota(b) = \frac{a}{1}\frac{b}{1} = \frac{ab}{1} = \iota(ab).$$

It remains for us to show only that ι is one to one. If $\iota(a) = \iota(b)$, then $\frac{a}{1} = \frac{b}{1}$, so $(a, 1) \sim (b, 1)$ giving $a1 = 1b$; that is, $a = b$. Thus ι is an isomorphism of D with $\iota(D)$, of course, as a subdomain of F. $\square$

Since $\frac{a}{b} = \frac{a}{1}\frac{1}{b} = \frac{a}{1}(\frac{b}{1})^{-1} = \iota(a)\iota(b)^{-1}$ clearly holds in F, we have now proved the following theorem.

Theorem 1.3.6. *Any integral domain* D *can be enlarged to (or embedded in) a field* F *such that every element of* F *can be expressed as a quotient of two elements of* D. *(Such a field* F *is called a* **field of quotients of** D, *or* **field of fractions of** D.)

The next theorem will show that the field of quotients of D is unique.

Theorem 1.3.7. *Let* F *be a field of quotients of* D *and let* E *be any field containing* D. *Then there exists a map* $\psi : F \to E$ *that gives an isomorphism of* F *with a subfield of* E *such that* $\psi(a) = a$ *for* $a \in D$.

Proof. For $a, b \in D$, by a/b we mean the quotient regarded as elements of F, by $a/_E b$ we mean the quotient regarded as elements of E. Define

$$\psi : F \to E, \quad \psi(a/b) = a/_E b, \forall a, b \in D \text{ with } b \neq 0.$$

$$\begin{array}{ccc} D & \leq & F \\ \| & & \downarrow \psi \\ D & \leq & E \end{array}$$

We first show that ψ is well-defined. If $a/b = c/d$ in F, then $ad = bc$ in D. Thus

$$a/_E b = c/_E d,$$

in E, so ψ is well-defined. The equations

$$\psi(xy) = \psi(x)\psi(y), \text{ and } \psi(x + y) = \psi(x) + \psi(y), \forall x, y \in F$$

follow easily from the definition of ψ on F and from the fact that ψ is the identity on D.

If $a/_E b = c/_E d$ we have $ad = bc$. So $a/b = c/d$. Thus ψ is one to one.

By definition, $\psi(a) = a$ for $a \in D$. The theorem follows. □

Theorem 1.3.8. *Every field E containing an integral domain D contains a field of quotients of D.*

Proof. Let F be a field of quotients of D. In the above Theorem the subfield $\psi[F]$ of E is a quotient field of D. □

Theorem 1.3.9. *Any two fields of quotients of an integral domain D are isomorphic.*

Proof. This directly follows from Theorem 1.3.7. □

We remark that, in general, not every unital noncommutative ring without zero divisors can be embedded into a division ring. This leads to Ore theory.

The **right Ore condition** for a multiplicative subset $S = R \setminus \{0\}$ of a ring R is that for any $a \in R$ and any $s \in S$, the intersection $aS \cap sR \neq \emptyset$. A (non-commutative) integral domain for which the set of non-zero elements satisfies the right Ore condition is called a **right Ore domain**. Only right Ore domains can be embedded in some division rings.

1.4. Rings of polynomials.

Definition 1.4.1. *Let R be a ring. A **polynomial** $f(x)$ with coefficients in R is a sum*

$$f(x) = a_0 + a_1 x + \cdots + a_n x^n$$

*where $n \in \mathbb{Z}^+, a_i \in R$. The a_i's are **coefficients** of $f(x)$. If $a_n \neq 0$ then a_n is called the **leading coefficient** of $f(x)$. A polynomial $f(x) \in R[x]$ is called a **monic polynomial** if its leading coefficient is 1 (assuming that R is unital).*

If R is unital, we will write a term $1x^k$ in such a sum as x^k. For example, in $\mathbb{Z}[x]$, we will write the polynomial $2 + 1x$ as $2 + x$. Finally, we shall agree that we may omit altogether from the formal sum any term $0x^i$, or a_0 if $a_0 = 0$ but not all $a_i = 0$. Thus $0, 2, x$, and $2 + x^2$

are polynomials with coefficients in $\mathbb{Z}$. An element of R is a constant polynomial.

Addition and multiplication of polynomials with coefficients in a ring R are defined in a way familiar to us. If

$$f(x) = a_0 + a_1 x + \cdots + a_n x^n,$$

and

$$g(x) = b_0 + b_1 x + \cdots + b_n x^n,$$

then for polynomial addition, we have

$$f(x) + g(x) = (a_0 + b_0) + (a_1 + b_1)x + \cdots + (a_n + b_n)x^n,$$

and for polynomial multiplication, we have

$$f(x)g(x) = d_0 + d_1 x + \cdots + d_{2n}x^{2n} \text{ where } d_k = \sum_{i=0}^{k} a_i b_{k-i}.$$

With these definitions of addition and multiplication, we have the following theorem.

Theorem 1.4.2. *The set $R[x]$ of all polynomials in an indeterminate x with coefficients in a ring R is a ring under polynomial addition and multiplication. If R is commutative, then so is $R[x]$, and if R unital then so is $R[x]$.*

Proof. Clearly, we have the abelian group $(R[x], +, 0)$. The associative law for multiplication and the distributive laws are straightforward, but slightly cumbersome, computations. We only prove the associative law.

Applying ring axioms to $a_i, b_j, c_k \in R$, we obtain

$$\left[\left(\sum_{i=0}^{m} a_i x^i \right) \left(\sum_{j=0}^{n} b_j x^j \right) \right] \left(\sum_{k=0}^{r} c_k x^k \right) = \sum_{s=0}^{m+n+r} \left(\sum_{i+j+k=s} a_i b_j c_k \right) x^s$$

$$= \left(\sum_{i=0}^{m} a_i x^i \right) \left[\left(\sum_{j=0}^{n} b_j x^j \right) \left(\sum_{k=0}^{r} c_k x^k \right) \right].$$

The distributive laws are similarly proved.

The comments prior to the statement of the theorem show that $R[x]$ is a commutative ring if R is commutative, and a unity $1 \neq 0$ in R is also unity for $R[x]$, in view of the definition of multiplication in $R[x]$. $\qquad\square$

Thus $\mathbb{Z}[x]$ is the ring of polynomials in the indeterminate x with integral coefficients, $\mathbb{Q}[x]$ the ring of polynomials in x with rational coefficients, and so on.

Example 1.4.1. *In $\mathbb{Z}_2[x]$, we have*

$$(x+1)^2 = (x+1)(x+1) = x^2 + (1+1)x + 1 = x^2 + 1.$$

Still working in $\mathbb{Z}_2[x]$, we obtain

$$(x+1) + (x+1) = (1+1)x + (1+1) = 0x + 0 = 0.$$

If R is a ring and x and y are two indeterminates, then we can form the ring $(R[x])[y]$, that is, the ring of polynomials in y with coefficients that are polynomials in x. Every polynomial in y with coefficients that are polynomials in x can be rewritten in a natural way as a polynomial in x with coefficients that are polynomials in y. This indicates that $(R[x])[y]$ is naturally isomorphic to $(R[y])[x]$, although a careful proof is tedious. We shall identify these rings by means of this natural isomorphism, and shall consider this ring $R[x, y]$ the ring of polynomials in two indeterminates x and y with coefficients in R. The ring $R[x_1, \cdots, x_n]$ of polynomials in the n indeterminates x_i with coefficients in R is similarly defined.

(a). If D is an integral domain then so is $D[x]$. In particular, if F is a field, then $F[x]$ is an integral domain.

(b). We can construct the field of quotients $F(x)$ of $F[x]$. Any element in $F(x)$ can be represented as a quotient $f(x)/g(x)$ of two polynomials in $F[x]$ with $g(x) \neq 0$. We similarly define $F(x_1, \cdots, x_n)$ to be the field of quotients of $F[x_1, \cdots, x_n]$. This field $F(x_1, \cdots, x_n)$ is the field of rational functions in n indeterminates over F. These fields play a very important role in algebraic geometry.

Let E and F be fields, with F a subfield of E, that is, $F \leq E$. The next theorem asserts the existence of very important homomorphisms of $F[x]$ into E. These homomorphisms will be the fundamental tools in this book.

Theorem 1.4.3. *Let F be a subfield of a field E, let $\alpha \in E$, and let x be an indeterminate. The map $\phi_\alpha : F[x] \to E$ defined by*

$$\phi_\alpha(a_0 + a_1 x + \cdots + a_n x^n) = a_0 + a_1 \alpha + \cdots + a_n \alpha^n$$

is a ring homomorphism of $F[x]$ into E. Also, $\phi_\alpha(x) = \alpha$, and $\phi_\alpha(a) = a$ for any $a \in F$. The homomorphism ϕ_α is called the **evaluation** *at α.*

Proof. Clearly ϕ_α is well defined. If

$$f(x) = a_0 + a_1 x + \cdots + a_n x^n, g(x) = b_0 + b_1 x + \cdots + b_m x^m,$$

$$h(x) = f(x) + g(x) = c_0 + c_1 x + \cdots + c_r x^r,$$

then

$$\phi_\alpha(f(x) + g(x)) = \phi_\alpha(h(x)) = c_0 + c_1 \alpha + \cdots + c_r \alpha^r,$$

while

$$\phi_\alpha(f(x)) + \phi_\alpha(g(x)) = (a_0 + a_1\alpha + \cdots + a_n\alpha^n) + (b_0 + b_1\alpha + \cdots + b_m\alpha^m).$$

Since by definition of polynomial addition we have $c_i = a_i + b_i$, we see that

$$\phi_\alpha(f(x) + g(x)) = \phi_\alpha(f(x)) + \phi_\alpha(g(x)).$$

For the multiplication, if $f(x)g(x) = d_0 + d_1 x + \cdots + d_s x^s$, then

$$\phi_\alpha(f(x)g(x)) = d_0 + d_1\alpha + \cdots + d_s\alpha^s,$$

while

$$(\phi_\alpha(f(x)))(\phi_\alpha(g(x))) = (a_0 + a_1\alpha + \cdots + a_n\alpha^n)(b_0 + b_1\alpha + \cdots + b_m\alpha^m).$$

Thus ϕ_α is a homomorphism.

The definition of ϕ_α applied to a constant polynomial $a \in F[x]$, where $a \in F$, gives $\phi_\alpha(a) = a$. Clearly,

$$\phi_\alpha(x) = \phi_\alpha(1x) = 1\alpha = \alpha.$$

$\square$

Example 1.4.2.   *(a). Consider the evaluation homomorphism $\phi_0 : \mathbb{Q}[x] \to \mathbb{R}$. Here*

$$\phi_0(a_0 + a_1 x + \cdots + a_n x^n) = a_0 + a_1 0 + \cdots + a_n 0^n = a_0.$$

Thus every polynomial is mapped onto its constant term.

(b). We have the evaluation homomorphism $\phi_2 : \mathbb{Q}[x] \to \mathbb{R}$. Here

$$\phi_2(a_0 + a_1 x + \cdots + a_n x^n) = a_0 + a_1 2 + \cdots + a_n 2^n.$$

Note that $\phi_2(x^2 + x - 6) = 2^2 + 2 - 6 = 0$. Thus $x^2 + x - 6$ is in the kernel N of ϕ_2. Of course, $x^2 + x - 6 = (x - 2)(x + 3)$, and the reason that $\phi_2(x^2 + x - 6) = 0$ is that $\phi_2(x - 2) = 2 - 2 = 0$.

(c). We have the evaluation homomorphism

$$\phi_i : \mathbb{Q}[x] \to \mathbb{C}, \quad \phi_i(a_0 + a_1 x + \cdots + a_n x^n) = a_0 + a_1 i + \cdots + a_n i^n$$

and $\phi_i(x) = i$. Note that $\phi_i(x^2 + 1) = i^2 + 1 = 0$, so $x^2 + 1$ is in the kernel N of ϕ_i.

(d). We have the evaluation homomorphism

$$\phi_\pi : \mathbb{Q}[x] \to \mathbb{R}, \quad \phi_\pi(a_0 + a_1 x + \cdots + a_n x^n) = a_0 + a_1 \pi + \cdots + a_n \pi^n.$$

We know that $a_0 + a_1 \pi + \cdots + a^n \pi^n = 0$ if and only if $a_i = 0$ for $i = 0, 1, \cdots, n$. Thus the kernel of ϕ_π is 0, and ϕ_π is a one-to-one map. This shows that all formal polynomials in π with rational coefficients form a ring isomorphic to $\mathbb{Q}[x]$ in a natural way with $\phi_\pi(x) = \pi$.

Using evaluation homomorphisms, by solving a polynomial equation, we shall refer to finding a zero of a polynomial.

Definition 1.4.4. *Let F be a subfield of a field E, and let $\alpha \in E$. Let $f(x) = a_0 + a_1 x + \cdots + a_n x^n \in F[x]$. Let $f(\alpha)$ denote*

$$\phi_\alpha(f(x)) = a_0 + a_1 \alpha + \cdots + a_n \alpha^n.$$

*If $f(\alpha) = 0$, then α is a **zero** of $f(x)$.*

In terms of this definition, we can rephrase the classical problem of finding all real numbers r such that $r^2 + r - 6 = 0$ by

$$\{\alpha \in \mathbb{R} : \phi_\alpha(x^2 + x - 6) = 0\} = \{r \in \mathbb{R} : r^2 + r - 6 = 0\} = \{2, -3\}.$$

1.5. Ideal theory.

Let R be a ring. We know that the two trivial ideals of R are R itself and 0. If $I \trianglelefteq R$ and $I \neq R$, then I is said to be a **proper ideal** of R, denoted $I \trianglelefteq R$. Let's introduce the following result.

Theorem 1.5.1. *Let R be a unital ring and let $I \trianglelefteq R$ such that I contains a unit. Then $I = R$.*

Proof. Let $u \in I$ be a unit. There exists $u^{-1} \in R$ such that

$$1 = u u^{-1} \in I.$$

Then $R = R \cdot 1 \subseteq I \subseteq R$, yielding that $R = I$. $\qquad \square$

Let's define what a maximal ideal is.

Definition 1.5.2. *A* **maximal ideal** *of a ring R is a proper ideal M of R so that there is no proper ideal of R properly containing M. In other words, a proper ideal M of R is maximal if*

$$M \subseteq N \trianglelefteq R \Longrightarrow N = M \text{ or } N = R.$$

Example 1.5.1. *Let $R = (\mathbb{Z}, +, \cdot)$ and p be a prime. Show that $p\mathbb{Z} \trianglelefteq R$, and $p\mathbb{Z}$ is a maximal ideal of $\mathbb{Z}$.*

Proof. It is easy to see that $p\mathbb{Z} \trianglelefteq R$.

Let $p\mathbb{Z} \subseteq N \trianglelefteq \mathbb{Z}$. If $p\mathbb{Z} \neq N$, there is $r \in N \setminus p\mathbb{Z}$. We see that $p \nmid r$, and $\gcd(p, r) = 1$. there exist $a, b \in \mathbb{Z}$ so that $1 = ap + br \in N$. From Theorem 1.5.1 we see that $N = \mathbb{Z}$. So $p\mathbb{Z}$ is a maximal ideal. □

Lemma 1.5.3. *Let R be a unital commutative ring. Then R is a field if and only if R has exactly the two ideals, i.e., 0 and R.*

Proof.

$$\begin{aligned}
R \text{ is a field} \quad &\Longleftrightarrow \quad \text{any nonzero element of } R \text{ is a unit}\\
&\Longleftrightarrow \quad \text{any nonzero ideal contains a unit}\\
&\Longleftrightarrow \quad \text{any nonzero ideal is } R\\
&\Longleftrightarrow \quad R \text{ has only the trivial ideals.}
\end{aligned}$$

□

Theorem 1.5.4. *Let R be a unital commutative ring and $M \trianglelefteq R$. Then M is maximal if and only if R/M is a field.*

Proof.

$$\begin{aligned}
M \text{ is max} \quad &\Longleftrightarrow \quad \{M, R\} = \{N \trianglelefteq R \mid N \supseteq M\}\\
&\Longleftrightarrow \quad \{N' \trianglelefteq R/M\} = \{M/M, R/M\} = \{0, R/M\}\\
&\Longleftrightarrow \quad R/M \text{ has exactly two ideals}\\
&\Longleftrightarrow \quad R/M \text{ is a field.}
\end{aligned}$$

□

Definition 1.5.5. *Let R be a commutative ring, $N \trianglelefteq R$. We say that N is* **prime** *if $ab \in N$ implies that $a \in N$ or $b \in N$ for $a, b \in R$. The set of all prime ideals of R is call the* **spectrum** *of R.*

Example 1.5.2. *A non-zero ideal $n\mathbb{Z} \trianglelefteq \mathbb{Z}$ is prime if and only if n is a prime.*

Proof. Suppose $n = p$ is a prime, and $a \cdot b \in p\mathbb{Z}$. So $p \mid a \cdot b$. So $p \mid a$ or $p \mid b$, i.e., $a \in p\mathbb{Z}$ or $b \in p\mathbb{Z}$.

For the other direction, suppose $n = pq$ is a composite number $(p, q \neq 1)$. Then $n \in n\mathbb{Z}$ but $p \notin n\mathbb{Z}$ and $q \notin n\mathbb{Z}$, since $0 < p, q < n$. $\square$

Definition 1.5.6. *Let R be a ring, $S \subseteq R$. The ideal $\langle S \rangle$ is the smallest ideal of R containing S, which is called the **ideal generated by** S.*

For any $A, B \subseteq R$, we define

$$AB = \left\{ \sum_{i=1}^{n} a_i b_i : a_i \in A, b_i \in B \right\}.$$

If R is unital, then $\langle S \rangle = SR + RS$.

Theorem 1.5.7. *Let R be a unital commutative ring, $N \trianglelefteq R$. Then N is prime if and only if R/N is an ID.*

Proof. ($\Rightarrow$). Suppose that N is prime. R/N is a unital commutative ring where $1 \neq 0$. We need to demonstrate that there are no zero divisors. Let $a + N, b + N \in R/N$ with $(a + N)(b + N) = 0 + N$ (since either a or b could be zero divisors). Then

$$ab + N = N \Rightarrow ab \in N.$$

Since N is prime, it follows that $a \in N$ or $b \in N$. So $a + N = 0 + N$ or $b + N = 0 + N$. So R/N has no zero divisors. By definition, it follows that R/N is an ID.

($\Leftarrow$). Suppose that the quotient ring R/N is an ID. Let $ab \in N$ for $a, b \in R$. Then

$$(a + N)(b + N) = ab + N = 0 + N \text{ in } R/N.$$

Since R/N is an ID, then either $a + N = 0 + N$ or $b + N = 0 + N$. So $a \in N$ or $b \in N$. So N is prime. $\square$

Combining this theorem with the fact that any field is an integral domain, we have the following corollary.

Corollary 1.5.8. *Let R be a unital commutative ring and $M \trianglelefteq R$. Then any maximal ideal of R is prime.*

Example 1.5.3. *Is the ideal $\langle 2, x^2 + 5 \rangle \subset \mathbb{Z}[x]$ prime?*

Solution. Consider $\mathbb{Z}[x]/\langle 2, x^2 + 5\rangle \cong \mathbb{Z}_2[x]/\langle x^2 - 1\rangle$. Then $x + 1, x - 1 \neq 0$ in $\mathbb{Z}_2[x]/\langle x^2 - 1\rangle$, but $(x + 1)(x - 1) = x^2 - 1 = 0$. So $\mathbb{Z}[x]/\langle 2, x^2 + 5\rangle$ is not an ID. Thus the ideal $\langle 2, x^2 + 5\rangle$ is not a prime ideal of $\mathbb{Z}[x]$. □

Lemma 1.5.9. *Let R be a finite commutative unital ring. Then any prime ideal of R is maximal.*

Proof. Let $N \trianglelefteq R$ and N be prime. We show that N is maximal. So R/N is an integral domain (by the previous theorem). Since R is finite, then R/N is finite. Since any finite integral domain is a field, then R/N is a field and so, N must be maximal. □

Definition 1.5.10. *If R is a commutative unital ring and $a \in R$, the ideal $\{ra|r \in R\}$ of all multiples of a is the **principal ideal** generated by a and is denoted by $\langle a\rangle$. An ideal N of R is **a principal ideal** if $N = \langle a\rangle$ for some $a \in R$. An integral domain is **a principal ideal domain** (PID, for short) if all its ideals are principal.*

Theorem 1.5.11. *For any field F, the polynomial ring $F[x]$ is a PID.*

Proof. Let $N \trianglelefteq F[x]$. If $N = \{0\}$, then $N = \langle 0\rangle$. Suppose that $N \neq \{0\}$, and let $g(x)$ be a nonzero element of N of minimal degree. If $\deg g(x) = 0$, then $g(x) \in F$ and is a unit. We see that $N = F[x] = \langle 1\rangle$, so N is principal. If $\deg g(x) \geq 1$, for any $f(x) \in N$, then

$$f(x) = g(x)q(x) + r(x), \text{ where } \deg r(x) < \deg g(x).$$

Since $f(x) \in N$ and $g(x) \in N$ then $f(x) - g(x)q(x) = r(x) \in N$ by definition of an ideal. Since $g(x)$ is a nonzero element of minimal degree in N, so $r(x) = 0$. Thus $f(x) = g(x)q(x) \in \langle g(x)\rangle$ and $N \subseteq \langle g(x)\rangle \subseteq N$. So $N = \langle g(x)\rangle$, i.e., $F[x]$ is a PID. □

Similarly, we can easily show the following result.

Corollary 1.5.12. *The ring $(\mathbb{Z}, +, \cdot)$ is a principal ideal domain.*

Theorem 1.5.13. *Let $p(x) \in F[x]$. Then $\langle p(x)\rangle$ is maximal in $F[x]$ if and only if $p(x)$ is irreducible over F.*

Proof. ($\Rightarrow$). Since $\langle p(x)\rangle$ is a maximal ideal of $F[x]$, then $\langle p(x)\rangle \neq \{0\}$ and $\langle p(x)\rangle \neq F[x]$, so $p(x) \notin F$. Let $p(x) = f(x)g(x)$ where

$f(x), g(x) \in F[x]$. Since $\langle p(x) \rangle$ is a maximal ideal and hence also a prime ideal, $f(x)g(x) \in \langle p(x) \rangle$ implies that

$$f(x) \in \langle p(x) \rangle \text{ or } g(x) \in \langle p(x) \rangle.$$

We may assume that $f(x) \in \langle p(x) \rangle$. Then $f(x) = p(x)u(x)$ for some $u(x) \in F[x]$. So $p(x)u(x)g(x) = p(x)$, i.e., $g(x)u(x) = 1$. Thus $g(x)$ is a unit in $F[x]$. So $p(x)$ is irreducible over F.

($\Leftarrow$). Suppose that $N \trianglelefteq F[x]$ such that $\langle p(x) \rangle \subseteq N \subseteq F[x]$. Since N is a principal ideal, we may assume that $N = \langle g(x) \rangle$ for some $g(x) \in N$. Then $p(x) \in N$ implies that

$$p(x) = g(x)q(x) \text{ for some } q(x) \in F[x].$$

Since $p(x)$ is irreducible, either $g(x)$ or $q(x)$ is of degree 0. If $g(x)$ is of degree 0, that is, a nonzero constant in F, then $g(x)$ is a unit in $F[x]$. Thus $\langle g(x) \rangle = N = F[x]$. If $q(x)$ is of degree 0, then $q(x) = c \in F$. So $g(x) = (1/c)p(x)$ is in $\langle p(x) \rangle$, i.e., $N = \langle p(x) \rangle$. Hence $\langle p(x) \rangle$ is maximal. $\square$

Theorem 1.5.14. *Let $p(x)$ be an irreducible polynomial in $F[x]$. If $p(x)|r(x)s(x)$ for $r(x), s(x) \in F[x]$, then either $p(x)|r(x)$ or $p(x)|s(x)$.*

Proof. Suppose $p(x)|r(x)s(x)$. Then $r(x)s(x) \in \langle p(x) \rangle$, which is maximal. Therefore, $\langle p(x) \rangle$ is a prime ideal. Hence $r(x)s(x) \in \langle p(x) \rangle$ implies that either $r(x) \in \langle p(x) \rangle$, yielding $p(x)|r(x)$, or that $s(x) \in \langle p(x) \rangle$, yielding $p(x)|s(x)$. $\square$

Example 1.5.4. *Suppose that D is an integral domain. If $D[x]$ is a principal ideal domain, show that D is a field.*

Proof. We need to show that any $a \in D \setminus \{0\}$ has an inverse. We have to use the given condition that $D[x]$ is a principal ideal domain. So consider the ideal $\langle a, x \rangle \trianglelefteq D[x]$. Since it is principal there is $f(x) \in D[x]$ such that $\langle a, x \rangle = \langle f(x) \rangle$. Since $f|a$ we know that $f = d$ for some $d \in D$. At the same time $d|x$. The $x = d(bx + c)$ for some $b, c \in D$. It follows that $bd = 1$, i.e., d is invertible. Thus $\langle a, x \rangle = D[x]$. So

$$1 = xg(x) + ah(x)$$

for some $g, h \in D[x]$. We have $1 = ah(0)$. So a is invertible. Therefore D is a field. $\square$

Example 1.5.5. *Prove that a prime ideal N in a commutative ring R contains every nilpotent element. Deduce that the **nilradical** of R*

(the set of all nilpotent elements in R) is contained in the intersection of all the prime ideals of R.

Proof. We first show by induction on n that $a \in N$ if $a^n \in N$ for some $n \in \mathbb{N}$. This is true for $n = 1$. If $n > 1$, since $a^n = aa^{n-1} \in N$ we see that $a^{n-1} \in N$. By inductive hypothesis we deduce that $a \in N$.

Suppose $b \in R$ is nilpotent, i.e., $b^n = 0$ for some $n \in \mathbb{N}$. Then $b^n \in N$. From the above established result we see that $b \in N$.   □

Solution. Let $x \in R$ with $x^n = 0$ for some $n \in \mathbb{N}$. Since N is a prime ideal of R, we know that R/N in an ID. From $(x + N)^n = N$ (the zero element in R/N), we see that $x + N = N$. Thus $x \in N$.   □

Example 1.5.6. *Find all prime ideals and all maximal ideals of the finite commutative ring $(\mathbb{Z}_{12}, +, \cdot)$.*

Solution. Note that $12 = 2^2 3$. We know that the only subgroups of the cyclic group $(\mathbb{Z}_{12}, +)$ are

$$\{0\}, 2\mathbb{Z}_{12}, 3\mathbb{Z}_{12}, 4\mathbb{Z}_{12}, 6\mathbb{Z}_{12}, \mathbb{Z}_{12}.$$

We can easily see that all of them are the ideals of $\mathbb{Z}_{12}$. So they are all the ideals of $\mathbb{Z}_{12}$. Note that the prime and the maximal ideals coincide for any finite unital commutative ring. The prime and maximal ideals are

$$2\mathbb{Z}_{12} = \{0, 2, 4, 6, 8, 10\} \text{ and } 3\mathbb{Z}_{12} = \{0, 3, 6, 9\}$$

because the factor rings are isomorphic to the fields

$$\frac{\mathbb{Z}_{12}}{2\mathbb{Z}_{12}} \cong \mathbb{Z}_2 \text{ and } \frac{\mathbb{Z}_{12}}{3\mathbb{Z}_{12}} \cong \mathbb{Z}_3$$

respectively, and any other quotients are not fields.   □

Example 1.5.7. *Let F and K be fields. If $F[x] \cong K[x]$, prove that $F \cong K$.*

Proof. Let $\varphi : F[x] \to K[x]$ be a ring isomorphism. Since $\mathcal{U}(K) = K^*$ and $\mathcal{U}(F) = F^*$, we deduce that $\varphi(F^*) = K^*$. Then $\varphi|_F : F \to K$ is an isomorphism of fields. So $F \cong K$.   □

1.6. Division algorithm for polynomials over a field.

In this section we always assume that F is field. The following theorem is the basic tool for our work in this section.

Theorem 1.6.1 (Division Algorithm). *Let* $f(x), g(x) \in F[x]$ *with* $g(x) \neq 0$. *Then there are unique polynomials* $q(x), r(x) \in F[x]$ *such that*

$$f(x) = g(x)q(x) + r(x), \text{ with } \deg(r(x)) < \deg(g(x)).$$

Proof. Let

$$f(x) = a_n x^n + a_{n-1} x^{n-1} + \cdots + a_0,$$

$$g(x) = b_m x^m + b_{m-1} x^{m-1} + \cdots + b_0 \neq 0,$$

where $b_m \neq 0$. Let

$$S = \{f(x) - g(x)s(x) : s(x) \in F[x]\}.$$

If $0 \in S$ then there exists an $s(x)$ such that $f(x) - g(x)s(x) = 0$, so $f(x) = g(x)s(x)$. Taking $q(x) = s(x)$ and $r(x) = 0$, we are done. Otherwise, let $r(x)$ be an element of minimal degree in S. Then

$$f(x) = g(x)q(x) + r(x)$$

for some $q(x) \in F[x]$. We must show that the degree of $r(x)$ is less than m. Suppose that

$$r(x) = c_t x^t + c_{t-1} x^{t-1} + \cdots + c_0,$$

with $c_j \in F$ and $c_t \neq 0$. If $t \geq m$, then

$$f(x) - q(x)g(x) - (c_t/b_m)x^{t-m}g(x) = r(x) - (c_t/b_m)x^{t-m}g(x),$$

and the latter is of the form $r_1(x) = r(x) - (c_t x^t + \text{terms of lower degree})$, which is a polynomial of degree lower than t, the degree of $r(x)$. However, the polynomial can be written in the form

$$r_1(x) = f(x) - g(x)[q(x) + (c_t/b_m)x^{t-m}],$$

so $r_1(x) \in S$, contradicting the fact that $r(x)$ was selected to have minimal degree in S. Thus the degree of $r(x)$ is less than the degree m of $g(x)$.

For uniqueness, suppose

$$f(x) = g(x)q_1(x) + r_1(x) = g(x)q_2(x) + r_2(x).$$

We see that $g(x)[q_1(x) - q_2(x)] = r_2(x) - r_1(x)$. Because $\deg(r_2(x) - r_1(x)) < m$, this can hold only if $q_1(x) - q_2(x) = 0$ so $q_1(x) = q_2(x)$. Then we must have $r_2(x) - r_1(x) = 0$ so $r_1(x) = r_2(x)$. $\square$

We can compute the polynomials $q(x)$ and $r(x)$ of the above theorem by long division just as we divided polynomials in $\mathbb{R}[x]$ in high school.

Example 1.6.1. *Let us work with polynomials in $\mathbb{Z}_5[x]$ and divide*

$$f(x) = x^4 - 3x^3 + 2x^2 + 4x - 1$$

by $g(x) = x^2 - 2x + 3$ to find q(x) and r(x) of the theorem.

Solution. The long division should be easy to follow, but remember that we are working in $\mathbb{Z}_5[x]$. For example, $4x - (-3x) = 2x$.

$$
\begin{array}{r}
x^2 \;\;- x - 3 \\
x^2 - 2x + 3 {\overline{\smash{\big)}\,x^4 - 3x^3 + 2x^2 + 4x - 1}} \\
\underline{-\,x^4 + 2x^3 - 3x^2} \\
-\,x^3 \;\;- x^2 + 4x \\
\underline{x^3 - 2x^2 + 3x} \\
-\,3x^2 + 7x - 1 \\
\underline{3x^2 - 6x + 9} \\
x + 8
\end{array}
$$

Thus $q(x) = x^2 - x - 3$, and $r(x) = x + 3$. $\square$

Theorem 1.6.2 (Factor Theorem). *Let $f(x) \in F[x]$ and $a \in F$. Then $f(a) = 0$ if and only if $x - a | f(x)$.*

Proof. From Theorem 1.6.1, we know that there are $q(x) \in F[x]$ and $r \in F$ such that

$$f(x) = q(x)(x - a) + r.$$

Then $f(a) = 0$ if and only if $r = 0$, if and only if $f(x) = q(x)(x - a)$, if and only if $x - a | f(x)$. $\square$

Example 1.6.2. *Again in $\mathbb{Z}_5[x]$, factor $x^4 + 3x^3 + 2x + 4$.*

Solution. Note that 1 is a zero of $x^4 + 3x^3 + 2x + 4$. We should be able to factor $x^4 + 3x^3 + 2x + 4$ into $(x - 1)q(x)$ in $\mathbb{Z}_5[x]$. Let us find the factorization by long division.

$$
\begin{array}{r}
x^3 + 4x^2 + 4x + 6 \\
x - 1{\overline{)\;\;x^4 + 3x^3 \qquad\quad + 2x + 4\;}} \\
-\,x^4 + x^3 \\ \hline
4x^3 \\
-\,4x^3 + 4x^2 \\ \hline
4x^2 + 2x \\
-\,4x^2 + 4x \\ \hline
6x + 4 \\
-\,6x + 6 \\ \hline
10
\end{array}
$$

Thus $x^4 + 3x^3 + 2x + 4 = (x - 1)(x^3 - x^2 - x + 1)$ in $\mathbb{Z}_5[x]$. Since 1 is seen to be a zero of $x^3 - x^2 - x + 1$ also, we can divide this polynomial by $x - 1$ and get

$$
\begin{array}{r}
x^2 \qquad\quad - 1 \\
x - 1{\overline{)\;\;x^3 - x^2 - x + 1\;}} \\
-\,x^3 + x^2 \\ \hline
-\,x + 1 \\
x - 1 \\ \hline
0
\end{array}
$$

Thus $x^4 + 3x^3 + 2x + 4 = (x - 1)^2(x^2 - 1) = (x - 1)^3(x + 1)$ in $\mathbb{Z}_5[x]$. $\qquad\square$

The next corollary should also look familiar.

Corollary 1.6.3. *Any $f(x) \in F[x]$ with $\deg(f(x)) = n \geq 1$ has at most n distinct zeros in F.*

Proof. We will prove this by induction on n. For $n = 1$, the statement is clear. Now suppose that the statement holds for $n = k \geq 1$ and we consider $n = k + 1$. If $f(x)$ does not have any zeros, the statement holds clearly. If $a_1 \in F$ is a zero of $f(x)$, then $f(x) = (x - a_1)g(x)$, where $g(x) \in F[x]$ with $\deg(g(x)) = k$. A zero $a_2 \in F$ of $f(x)$ with $a_2 \neq a_1$ satisfies $(a_2 - a_1)g(a_2) = 0$, yielding that $g(a_2) = 0$, i.e., a_2 is a zero of $g(x)$. Since $g(x)$ has at most k zeros, we conclude that $f(x)$ has at most $k + 1$ zeros in F. The proof completes. $\qquad\square$

Let us recall a result on finitely generated abelian groups [ZTL, Theorem 4.1.11].

Theorem 1.6.4. *Any finitely generated abelian group is isomorphic to*

$$\mathbb{Z}_{r_1} \times \mathbb{Z}_{r_2} \times \cdots \times \mathbb{Z}_{r_s} \times \mathbb{Z}^n$$

where $s, n \in \mathbb{Z}^+, r_i \in \mathbb{N}$ *and* $r_1 | r_2, r_2 | r_3, \cdots, r_{s-1} | r_s$.

Our next corollary is concerned with the structure of the multiplicative group F^* of nonzero elements of a field F, rather than with factorization in $F[x]$.

Corollary 1.6.5. *If G is a finite subgroup of the multiplicative group* $(F^*, \cdot, 1)$ *of a field F, then G is cyclic. In particular, the multiplicative group of all nonzero elements of a finite field is cyclic.*

Proof. As a finite abelian group, G is isomorphic to a direct product $C_{d_1} \times C_{d_2} \times \cdots \times C_{d_r}$, where each $d_i | d_{i+1}$, and each of the C_{d_i} is a cyclic group of order d_i in multiplicative notation. If $a_i \in C_{d_i}$, then $a_i^{d_i} = 1$, so $a_i^{d_r} = 1$ since d_i divides d_r. Thus for all $\alpha \in G$, we have $\alpha^{d_r} = 1$, so every element of G is zero of $x^{d_r} - 1$. But G has $d_1 d_2 \cdots d_r$ elements, while $x^{d_r} - 1$ can have at most d_r zeros in the field F, so $d_1 d_2 \cdots d_r \leq d_r$, we deduce that $d_1 = d_2 = \cdots = d_{r-1} = 1$. Therefore G is isomorphic to the cyclic group C_{d_r}.   □

Corollary 1.6.6. *If p is a prime, then*

(a). $n^{p-1} \equiv 1 \pmod{p}$ *for any $n \in \mathbb{Z}$ with $p \nmid n$;*
(b). $n^p \equiv n \pmod{p}$ *for any $n \in \mathbb{Z}$.*

Proof. Applying the above corollary we obtain the statements.
                    □

Example 1.6.3. *Find all zeros in $\mathbb{Z}_5$ for the polynomial* $2x^{219} + 3x^{74} + 2x^{57} + 3x^{44}$.

Solution. We can write $2x^{219} + 3x^{74} + 2x^{57} + 3x^{44} = 2(x^{219} - x^{74} + x^{57} - x^{44})$. Clearly 0 is a solution for $f(x) = x^{219} - x^{74} + x^{57} - x^{44}$. If $a \in \mathbb{Z}_5 \setminus \{0\}$, we see that $a^4 = 1$ in $\mathbb{Z}_5$. Then

$$f(a) = a^3 - a^2 + a - 1 = (a-1)(a^2+1) = (a-1)(a^2 - 2^2)$$
$$= (a-1)(a+2)(a-2).$$

Thus all zeros for the polynomial are $0, 1, 2, -2$.   □

1.7. Irreducible polynomials over a field.

We consider polynomials over a field F in this section.

Definition 1.7.1. *A nonconstant polynomial $f(x) \in F[x]$ is* **irreducible** *over F or is an* **irreducible polynomial** *in $F[x]$ if $f(x)$ cannot be expressed as a product $g(x)h(x)$ of two polynomials $g(x)$ and $h(x)$ in $F[x]$ both of lower degree than $\deg(f(x))$.*

This definition concerns the concept irreducible over F and not just the concept irreducible. A polynomial $f(x)$ may be irreducible over F, but may not be irreducible if viewed over a larger field E containing F.

Example 1.7.1. *We know that $x^2 - 3$ viewed in $\mathbb{Q}[x]$ has no zeros in $\mathbb{Q}$. This shows that $x^2 - 3$ is irreducible over $\mathbb{Q}$, for a factorization $x^2 - 3 = (ax + b)(cx + d)$ for $a, b, c, d \in \mathbb{Q}$ would give rise to zeros of $x^2 - 3$ in $\mathbb{Q}$. However, $x^2 - 3$ viewed in $\mathbb{R}[x]$ is not irreducible over $\mathbb{R}$, because $x^2 - 3$ factors in $\mathbb{R}[x]$ into $(x - \sqrt{3})(x + \sqrt{3})$.*

Note that the units in $F[x]$ are precisely the nonzero elements of F, i.e., $\mathcal{U}(F[x]) = F^*$. Thus we could have defined an irreducible polynomial $f(x)$ as a nonconstant polynomial such that in any factorization $f(x) = g(x)h(x)$ in $F[x]$, either $g(x)$ or $h(x)$ is a unit.

Irreducible polynomials will play a very important role in almost everywhere. The problem of determining whether a given $f(x) \in F[x]$ is irreducible over F may be difficult. We now give some criteria for irreducibility that are useful in certain cases.

Theorem 1.7.2. *Let $f(x) \in F[x]$, and let $f(x)$ be of degree 2 or 3. Then $f(x)$ is reducible over F if and only if it has a zero in F.*

Proof. "$\Leftarrow$". Suppose that $f(x)$ has a root a in F. By Theorem 1.6.2 we know that $x - a | f(x)$, i.e., $f(x) = (x - a)g(x)$ for some $g \in F[x]$ with positive degree since $\deg(f(x)) = 2$ or 3. Thus $f(x)$ is reducible.

"$\Rightarrow$". Suppose that $f(x)$ is reducible. Then $f(x) = g(x)h(x)$ where $g(x), h(x) \in F[x]$ of positive degrees. Let $\deg(g(x)) = r$ or $\deg(h(x)) = s$. We see that $r \geq 1, s \geq 1$ and $r + s = 2$ or 3 since $\deg(f(x)) = 2$ or 3. We deduce that $r = 1$ or $s = 1$. We may assume that $r = 1$, i.e., $g(x) = ax - b$ where $a \in F^*, b \in F$. Thus $g(x)$ has a root in $b/a \in F$, and hence $f(b/a) = 0$. $\qquad\square$

Example 1.7.2. *Show that $f(x) = x^3 + 3x + 2 \in \mathbb{Z}_5[x]$ is irreducible over $\mathbb{Z}_5$.*

Solution. We compute that

$$f(0) = 2, f(1) = 1, f(-1) = -2, f(2) = 1, \text{ and } f(-2) = -2,$$

showing that $f(x)$ has no zeros in $\mathbb{Z}_5$. From Theorem 1.7.2 we know that $f(x)$ is irreducible over $\mathbb{Z}_5$. □

We turn to some conditions for irreducibility of polynomials in $\mathbb{Q}[x]$.

Definition 1.7.3. $f(x) = \sum_0^n a_i x^i \in \mathbb{Z}[x]$ is **primitive** if $\deg f > 0$ and the coefficients $a_0, \ldots, a_n$ are relatively prime.

Lemma 1.7.4 (Gauss' Lemma). *If $f(x), g(x) \in \mathbb{Z}[x]$ are primitive, then so is $f(x)g(x)$.*

Proof. Let $f(x) = \sum_{i=0}^m a_i x^i$, $g(x) = \sum_{i=0}^n b_i x^i$, and $f(x)g(x) = \sum c_l x^l$, where $c_l = \sum_{i+j=l} a_i b_j$. Let $p \in \mathbb{Z}$ be prime. It suffices to show that there exists l such that $p \nmid c_l$. Suppose

$$s = \min\{i : p \nmid a_i\} \text{ and } t = \min\{i : p \nmid b_i\}.$$

Then

$$c_{s+t} = a_s b_t + \sum_{\substack{i+j=s+t \\ i<s}} a_i b_j + \sum_{\substack{i+j=s+t \\ j<t}} a_i b_j,$$

so $p \nmid c_{s+t}$ since $p \nmid a_s b_t$ but p divides all the other terms on the right hand side. So $f(x)g(x)$ is primitive. □

Lemma 1.7.5. *Let $f(x) \in \mathbb{Q}[x] \setminus \{0\}$. Then*

(1). there exists $c \in \mathbb{Q}$ and primitive $g(x) \in \mathbb{Z}[x]$ such that $f(x) = cg(x)$;

(2). if also $f(x) = dh(x)$ with $d \in \mathbb{Q}$ and primitive $h(x) \in \mathbb{Z}[x]$, then $c = \pm d$ and $g(x) = \pm h(x)$.

Proof. (i) Clearly there is $a \in \mathbb{Z}^*$ such that $af(x) = \sum_0^n a_i x^i \in \mathbb{Z}[x]$. Let $b = \gcd(a_0, \ldots, a_n)$. For each i put $b_i = a_i/b$, and then put $g(x) = \sum_0^n b_i x^i$. We have $f(x) = \frac{b}{a}g(x)$ by construction, and $g(x)$ is primitive since the b_i are relatively prime.

(ii) If also $f(x) = \frac{b'}{a'}h(x)$ with $a', b' \in \mathbb{Z}$ and primitive $h(x) \in \mathbb{Z}[x]$, then $a'bg(x) = ab'h(x) \in \mathbb{Z}[x]$. Since both $|a'b|$ and $|ab'|$ are the GCD of the coefficients of $a'bg(x) = ab'h(x)$, so $b/a = \pm b'/a'$, and $h(x) = \pm g(x)$. □

Lemma 1.7.6. *Let $f(x) \in \mathbb{Z}[x]$ be primitive. Then $f(x)$ is reducible in $\mathbb{Z}[x]$ if and only if $f(x)$ is reducible in $\mathbb{Q}[x]$.*

Proof. "⇐". Assume that $f(x)$ is reducible in $\mathbb{Q}[x]$, and choose $g(x), h(x) \in \mathbb{Q}[x]$ such that $f(x) = g(x)h(x)$ and $\deg g(x), \deg h(x) > 0$. From Lemma 1.7.5, there exist $a, b \in \mathbb{Q}^*$ that $ag(x), bh(x) \in \mathbb{Z}[x]$ are primitive. So $(ab)^{-1}(ag(x))(bh(x)) = f(x)$. Since both $(ag(x))(bh(x)), f(x)$ are primitive, from Lemma 1.7.5 we know that $f(x) = \pm(ag(x))(bh(x))$. Thus $f(x)$ is reducible in $\mathbb{Z}[x]$.

"⇒". Assume that $f(x)$ is reducible in $\mathbb{Z}[x]$, and choose nonconstant $g(x), h(x) \in \mathbb{Z}[x]$ such that $f(x) = g(x)h(x)$. Since $f(x)$ is primitive, then $g(x), h(x)$ are primitive, and $\deg g(x), \deg h(x) > 0$. Clearly $f(x)$ is reducible in $\mathbb{Q}[x]$. □

Theorem 1.7.7. *Let $f(x) \in \mathbb{Z}[x]$. Then $f(x)$ factors into a product of two polynomials of lower degrees r and s in $\mathbb{Q}[x]$ if and only if it has such a factorization with polynomials of the same degrees r and s in $\mathbb{Z}[x]$.*

Proof. "⇒". Suppose $f(x) = g(x)h(x)$ where $h(x), g(x) \in \mathbb{Q}[x]$ and $\deg g(x), \deg h(x) < \deg f(x)$. From Lemma 1.7.5, there exist $a, b \in \mathbb{Q}^*$ such that $ag(x), bh(x) \in \mathbb{Z}[x]$ are primitive, and there is $c \in \mathbb{Z}^*$ such that $f(x) = cf_1(x)$ where $f_1(x) \in \mathbb{Z}[x]$ is primitive. So $(ab)^{-1}(ag(x))(bh(x)) = cf_1(x)$. Since both $(ag(x))(bh(x)), f_1$ are primitive, from Lemma 1.7.5, we know that $f_1(x) = \pm(ag(x))(bh(x))$. Thus $f(x) = \pm c(ag(x))(bh(x))$ in $\mathbb{Z}[x]$.

"⇐". This is trivial since $\mathbb{Z}[x]$ is a subring of $\mathbb{Q}[x]$. □

Corollary 1.7.8. *Let $f(x) = a_0 + a_1 x + \cdots + a_n x^n \in \mathbb{Z}[x]$ with $n \geq 1$ and $a_0 a_n \neq 0$. If p/q, where $p, q \in \mathbb{Z}^*$ with $\gcd(p, q) = 1$, is a zero of $f(x)$, then $p|a_0, q|a_n$.*

Proof. From $f(p/q) = 0$ we see that

$$a_0 q^n + a_1 p q^{n-1} + \cdots + a_{n-1} p^{n-1} q + a_n p^n = 0.$$

So $p|a_0 q^n$ and $q|a_n p^n$. Since $\gcd(p, q) = 1$, then $p|a_0$ and $q|a_n$. □

Example 1.7.3. *(a). Is $\mathbb{Q}[x]/\langle x^2 - 6x - 6 \rangle$ a field? Why?*
(b). Find all $c \in \mathbb{Z}_5$ so that $\mathbb{Z}_5[x]/\langle x^2 + x + c \rangle$ is a field.

Solution. (a). We know that $\mathbb{Q}[x]/\langle x^2 - 6x - 6 \rangle$ is a field iff $\langle x^2 - 6x - 6 \rangle$ is maximal if and only if $x^2 - 6x - 6$ is irreducible.

Possible rational zeros of $x^2 - 6x - 6$ are $\pm 1, \pm 2, \pm 3, \pm 6$. By testing none of them are. So $x^2 - 6x - 6$ is irreducible over $\mathbb{Q}$. Thus $\mathbb{Q}[x]/\langle x^2 - 6x - 6 \rangle$ is a field.

(b). We know that $f(x) = x^2 + x + c$ is irreducible if and only if $f(x)$ has non zeros in $\mathbb{Z}_5$. Note that

$$f(0) = f(-1) = c, f(1) = f(-2) = c + 2, f(2) = c + 1.$$

Thus $f(x)$ with $c = 0, -1, -2$ is not irreducible. So $c = 1, 2$. $\square$

Example 1.7.4. *Show that $f(x) = x^4 + 8x^3 - 2x^2 + 1$ is irreducible over $\mathbb{Q}$.*

Solution. If $f(x)$ has a linear factor in $\mathbb{Q}[x]$, then it has a zero in $\mathbb{Z}$, and this zero would have to be a divisor in $\mathbb{Z}$ of 1, that is, either ± 1. But $f(1) = 8$, and $f(-1) = -8$, so such a factorization is impossible. If $f(x)$ factors into two quadratic factors in $\mathbb{Q}[x]$, then it has a factorization.

$$(x^2 + ax + b)(x^2 + cx + d)$$

in $\mathbb{Z}[x]$. Equating coefficients of powers of x, we find that we must have

$$bd = 1, ad + bc = 0, ac + b + d = -2, \text{ and } a + c = 8$$

for integers $a, b, c, d \in \mathbb{Z}$. From $bd = 1$, we see that either $b = d = 1$ or $b = d = -1$. In any case, $b = d$ and from $ad + bc = 0$, we deduce that $a + c = 0$ which is impossible. Thus a factorization into two quadratic polynomials is also impossible and $f(x)$ is irreducible over $\mathbb{Q}$. (We will have a much shorter proof for this in Example 1.8.1.) $\square$

No we provide the famous Schönemann-Eisenstein criterion for irreducibility.

Theorem 1.7.9 (Schönemann-Eisenstein Criterion). *Let $f(x) = a_n x^n + \cdots + a_1 x + a_0 \in \mathbb{Z}[x]$ with $n \geq 1$ and $a_n \neq 0$. If there is a prime p such that*

(i). $p \mid a_i$ for $0 \leq i < n$,
(ii). $p \nmid a_n$,
(iii). $p^2 \nmid a_0$,

then $f(x)$ is irreducible in $\mathbb{Q}[x]$.

Proof. To the contradiction, assume that $f(x)$ is reducible in $\mathbb{Q}[x]$. Then $f(x) = g(x)h(x)$ for some

$$g(x) = b_r x^r + \cdots + b_1 x + b_0 \in \mathbb{Q}[x] \setminus \mathbb{Q},$$
$$h(x) = c_s x^s + \cdots + c_1 x + c_0 \in \mathbb{Q}[x] \setminus \mathbb{Q}.$$

From Theorem 1.7.7, we may assume that $b_i, c_j \in \mathbb{Z}$. Denoting by $\overline{f}(x), \overline{g}(x), \overline{h}(x)$ the reductions mod p of these polynomials (i.e., we consider the coefficients as elements in $\mathbb{Z}_p$), we have $\overline{g}(x)\overline{h}(x) = \overline{f}(x) = \overline{a}_n x^n$, which means that $\overline{g}(x) = \overline{b}_r x^r$ and $\overline{h}(x) = \overline{c}_s x^s$. This shows that $p \mid b_0$ and $p \mid c_0$, yielding that $p^2 \mid b_0 c_0$, i.e., $p^2 \mid a_0$, which contradicts the hypothesis of the theorem. □

This criterion is named after Theodor Schönemann (1812–1868) and Gotthold Eisenstein (1823–1852). It was also known as the Eisenstein Criterion. Schönemann was the first to publish his proof in 1846, and then by Eisenstein in 1850.

Note that if we take $p = 2$, the Schönemann-Eisenstein Criterion gives us still another proof of the irreducibility of $x^2 - 2$ over $\mathbb{Q}$.

Example 1.7.5. *Taking $p = 3$ in the above theorem, we see that $25x^5 - 9x^4 - 3x^2 - 12$ is irreducible over $\mathbb{Q}$.*

Corollary 1.7.10. *The polynomial*

$$\Phi_p(x) = (x^p - 1)/(x - 1) = x^{p-1} + x^{p-2} + \cdots + x + 1$$

is irreducible over $\mathbb{Q}$ for any prime p.

Proof. It is clear that $\Phi_p(x)$ is irreducible over $\mathbb{Q}$ iff $\Phi_p(x + 1)$ is irreducible over $\mathbb{Q}$. Let

$$g(x) = \Phi_p(x+1) = ((x+1)^p - 1)/((x+1) - 1) = \frac{x^p + \binom{p}{1}x^{p-1} + \cdots + px}{x}.$$

The coefficient of x^{p-r} for $0 < r < p$ is the binomial coefficient $p!/(r!(p-r)!)$ which is divisible by p because $p \mid p!$ but $p \nmid r!$ or $p \nmid (p-r)!$ when $0 < r < p$. Thus

$$g(x) = x^{p-1} + \binom{p}{1}x^{p-2} + \cdots + p$$

satisfies the Schönemann-Eisenstein Criterion for the prime p and is thus irreducible over $\mathbb{Q}$. Thus $\Phi_p(x)$ must also be irreducible over $\mathbb{Q}$. □

The polynomial $\Phi_p(x)$ in the above corollary is called the **p-th cyclotomic polynomial**.

The p-th cyclotomic polynomial $\Phi_p(x)$ is generally not irreducible over a field of finite characteristic.

Example 1.7.6. *In $\mathbb{Z}_2[x]$,*

$$\Phi_7(x) = x^6 + x^5 + \cdots + x + 1 = (x^3 + x^2 + 1)(x^3 + x + 1).$$

Using Theorem 1.7.2 one can further show that both polynomials $x^3 + x^2 + 1$, $x^3 + x + 1$ are irreducible over $\mathbb{Z}_2$.

Theorem 1.7.11 (Unique Factorization Theorem). *For any $f(x) \in F[x]$ of positive degree, there exist a unique constant c; unique distinct irreducible monic polynomials $f_1(x), f_2(x), \cdots, f_k(x) \in F[x]$; and unique positive integers $n_1, n_2, \cdots, n_k$ such that*

$$f(x) = c f_1(x)^{n_1} f_2(x)^{n_2} \cdots f_k(x)^{n_k}.$$

Proof. Write $f(x)$ as a product of as many as possible positive polynomials

$$f(x) = g_1(x) g_2(x) \cdots g_r(x)$$

where each $g_i(x)$ is of positive degree and r is maximal. Then each $g_i(x)$ is irreducible. We may assume that $f(x) = c h_1(x) h_2(x) \cdots h_r(x)$, where each $h_i(x)$ is irreducible and monic.

Let us prove the uniqueness by induction on r. Suppose $f(x)$ has another decomposition of monic irreducible polynomials:

$$f(x) = c p_1(x) p_2(x) \cdots p_s(x),$$

where each $p_i(x)$ is irreducible and monic. If $r = 1$ the uniqueness is clear. Now assume that $r > 1$. From Theorem 1.5.14 there is a $p_j(x)$ such that $h_1(x) = p_j(x)$. Without loss of generality we may assume that $h_1(x) = p_1(x)$. The we obtain that

$$h_2(x) \cdots h_r(x) = p_2(x) \cdots p_s(x).$$

Using inductive hypothesis we deduce that $r = s$ and $h_i(x) = p_i(x)$ for all $i = 1, 2, \ldots, r$ after renumbering $p_i(x)$ if necessary. $\qquad \square$

Example 1.7.7. *By Schönemann-Eisenstein Criterion, we know that $x^4 - 4x^3 + 2x - 2$ is irreducible over $\mathbb{Q}$. But in $\mathbb{Z}_7[x]$,*

$$x^4 - 4x^3 + 2x - 2 = (x-2)^3(x+2) = (x-1)^2(2x-4)(4x+8).$$

Example 1.7.8. *Find primes p so that $x + 2 | x^4 + x^3 + x^2 - x + 1$ in $\mathbb{Z}_p[x]$.*

Solution. Let $f(x) = x^4 + x^3 + x^2 - x + 1$. From Factor Theorem 1.6.2 we know that $x + 2 | f(x)$ if and only if $f(-2) = 0$, i.e., $15 = 0$. Thus $p = 3$, or 5.

You may use Division Algorithm to solve this question. $\qquad \square$

It is generally hard to check whether a polynomial is irreducible over $\mathbb{Q}$. Here we provide a method to help do so.

Proposition 1.7.12. *Let $f(x) \in \mathbb{Z}[x]$. Suppose that $\overline{f}(x)\mathbb{Z}_p[x]$ is the reduction of $f(x)$ modulo p with $deg(\overline{f}) = deg(f)$. If $f(x)$ is reducible in $\mathbb{Q}[x]$, then $\overline{f}(x)$ is reducible in $\mathbb{Z}_p[x]$. Or by the contrapositive: if $\overline{f}(x)$ is irreducible in $\mathbb{Z}_p[x]$, then $f(x)$ is irreducible in $\mathbb{Q}[x]$.*

Example 1.7.9. *Show that the polynomial $f(x) = x^5 + (2a+1)x^2 + (2b+1) \in \mathbb{Z}[x]$ is irreducible for any integers $a, b \in \mathbb{Z}$.*

Proof. Consider the polynomial in $\mathbb{Z}_2[x]$. We have $\overline{f}(x) = x^5 + x^2 + 1$. Since $f(0)f(1) \neq 0$ we know that $\overline{f}(x)$ does not have degree one factors. If $\overline{f}(x)$ is not irreducible in $\mathbb{Z}_2[x]$, it must have an irreducible of degree 2 in $\mathbb{Z}_2[x]$. Since $x^2 + x + 1$ is the only irreducible of degree 2 in $\mathbb{Z}_2[x]$, and

$$\overline{f}(x) = (x^2 + x + 1)(x^3 + x^2) + 1,$$

we see that $x^2 + x + 1 \nmid \overline{f}(x)$. Thus $f(x)$ is irreducible for any integers $a, b \in \mathbb{Z}$. □

1.8. Other irreducibility criteria.
We will provide several other important simple-to-use irreducibility criteria for integer polynomials. The first one is the following Perron's irreducibility criterion which was first proved by Oskar Perron (1880–1975) in 1907 using Complex Analysis.

Theorem 1.8.1 (Perron's Irreducibility Criterion). *Suppose*

$$f(x) = x^n + a_{n-1}x^{n-1} + \cdots + a_1 x + a_0 \in \mathbb{Z}[x]$$

where $a_0 \neq 0$. If either of the following two conditions applies:

$$|a_{n-1}| > 1 + |a_{n-2}| + \cdots + |a_0|; \tag{1.2}$$

$$|a_{n-1}| = 1 + |a_{n-2}| + \cdots + |a_0| \text{ and } f(\pm 1) \neq 0, \tag{1.3}$$

then $f(x)$ is irreducible over $\mathbb{Q}$.

Let us first recall Rouche's Theorem from Complex Analysis.

Theorem 1.8.2 (Rouche's Theorem). *Let $f(z)$ and $g(z)$ be analytic functions on and inside a simple closed curve C. Suppose that $|f(z) + g(z)| < |f(z)| + |g(z)|$ for all points z on C. Then $f(z)$ and $g(z)$ have the same number of zeros (counting multiplicities) interior to C.*

Lemma 1.8.3. *Let $f(z) = z^n + a_{n-1}z^{n-1} + \cdots + a_1 z + a_0$ be as in Theorem 1.8.1. Then exactly one zero z of $f(x)$ satisfies $|z| > 1$, and the other $n - 1$ zeros of $f(z)$ satisfy $|z| < 1$.*

Proof. (a). Suppose that $|a_{n-1}| > 1 + |a_{n-2}| + \cdots + |a_0|$. We need to apply Rouche's Theorem to the functions $g(z) = z^n + a_{n-1}z^{n-1}$ and $f(z)$. For $|z| = 1$ we have

$$
\begin{aligned}
|f(z) - g(z)| &= |a_{n-2}z^{n-2} + \cdots + a_0| \\
&\leq |a_{n-2}| + \cdots + |a_0| < |a_{n-1}| - 1,
\end{aligned}
\tag{1.4}
$$

$$
|f(z)| + |g(z)| \geq |g(z)| = |z + a_{n-1}| \geq |a_{n-1}| - 1. \tag{1.5}
$$

It follows that $f(z)$ has the same number of zeros as $z^n + a_{n-1}z^{n-1}$ inside the unit circle. We know that $z^n + a_{n-1}z^{n-1}$ has $n - 1$ zeros inside the unit circle. It follows that $f(z)$ has exactly $n - 1$ zeros inside the unit circle. It is easy to see that $f(z)$ has no zeros on the unit circle.

(b). Suppose that $|a_{n-1}| = 1 + |a_{n-2}| + \cdots + |a_0|$ and $f(\pm 1) \neq 0$. Equation (1.4) becomes $|f(z) - g(z)| \leq |a_{n-1}| - 1$. Now in Equation (1.5), if the second inequality is not strict, then $z = \pm 1$ and the first inequality in Equation (1.5) is strict. Thus we also have $|f(z) - g(z)| < |f(z)| + |g(z)|$ on the unit circle. The remaining argument is the same as (a). □

Proof of Theorem 1.8.1. Suppose that $f(x) = g(x)h(x)$ where $g(x)$ and $h(x)$ are integer polynomials. Since, by the above Lemma, $f(x)$ has only one zero with modulus not less than 1, one of the polynomials $g(x), h(x)$ has all its zeroes strictly inside the unit circle. Suppose that $z_1, \ldots, z_k$ are the zeroes of $g(x)$, and $|z_1|, \ldots, |z_k| < 1$. Note that $g(0)$ is a nonzero integer, and $|g(0)| = |z_1 \cdots z_k| < 1$, contradiction. Therefore, $f(x)$ is irreducible. □

Example 1.8.1. *Let us revisit Example 1.7.4. Using Perron's Irreducibility Criterion, we easily see that the polynomial $f(x) = x^4 + 8x^3 - 2x^2 + 1$ is irreducible over $\mathbb{Q}$.*

It is interesting to know the following generalization. We will omit its proof.

Theorem 1.8.4 (Perron's Irreducibility Criterion). *Let F be a field. Suppose*

$$
f(x, y) = a_n(y)x^n + a_{n-1}(y)x^{n-1} + \cdots + a_1(y)x + a_0(y) \in F[x, y]
$$

where $a_i(y) \in F[y]$ with $a_0 a_n \neq 0$ and $a_n(y) \in F$. If

$$
\deg(a_{n-1}) > \max\{\deg(a_0), \deg(a_1), \cdots, \deg(a_{n-2})\}
$$

then $f(x, y)$ is irreducible over $F(y)$.

The other simple-to-use criterion is Cohn's irreducibility criterion which was first proved by Arthur Cohn (1894–1940) in 1925 using Number Theory. We omit its long proof here.

Theorem 1.8.5 (Cohn's Irreducibility Criterion). *Let*

$$f(x) = a_n x^n + a_{n-1} x^{n-1} + \cdots + a_1 x + a_0 \in \mathbb{Z}[x]$$

where all $a_i \geq 0$. Suppose that $b \in \mathbb{N}$ with $b \geq 2$ and $0 \leq a_i \leq b - 1$. If $f(b)$ is a prime, then $f(x)$ is irreducible over $\mathbb{Q}$.

We introduce our next simple-to-use criterion which was proved by Hiroyuki Osada in 1987.

Theorem 1.8.6 (Osada's Irreducibility Criterion). *Let p be a prime,*

$$f(x) = a_n x^n + a_{n-1} x^{n-1} + \cdots + a_1 x \pm p \in \mathbb{Z}[x],$$

with $n \geq 1$ and $a_n \neq 0$. If $p > |a_1| + \cdots + |a_n|$, then $f(x)$ is irreducible over $\mathbb{Q}$.

Proof. Let α be any complex zero of $f(x)$. Suppose that $|\alpha| \leq 1$, then $p = |a_1 \alpha + \cdots + a_n \alpha^n| \leq |a_1| + \cdots + |a_n|$, a contradiction. Therefore, all the zeros of $f(x)$ satisfies $|\alpha| > 1$. Now, suppose that $f(x) = g(x)h(x)$, where $g(x), h(x)$ are nonconstant integer polynomials. Then $a_0 = f(0) = g(0)h(0)$. Since p is prime, one of $|g(0)|, |h(0)|$ equals 1. Say $|g(0)| = 1$, and let b be the leading coefficient of $g(x)$. If $\alpha_1, \cdots, \alpha_k$ are the roots of $g(x)$, then $|\alpha_1 \alpha_2 \cdots \alpha_k| = 1/|b| \leq 1$. However, $\alpha_1, \cdots, \alpha_k$ are also zeros of $f(x)$, and so each has an magnitude greater than 1. Contradiction. Therefore, $f(x)$ is irreducible. $\square$

The last simple-to-use criterion is named after Alfred Brauer (1894–1985) who proved it in 1951.

Theorem 1.8.7 (Brauer's Irreducibility Criterion). *Let $a_1 \geq a_2 \geq \cdots \geq a_n$ be positive integers and $n \geq 2$. Then the polynomial $f(x) = x^n - a_1 x^{n-1} - \cdots - a_{n-1} x - a_n$ is irreducible over $\mathbb{Q}$.*

Proof. If all zeros of $f(x)$ are negative or non-real, a_n would be negative. So $f(x)$ has at least one non-negative zeros, say α. Then we see that

$$\alpha^n = a_1 \alpha^{n-1} + \cdots + a_{n-1} \alpha + a_n > a_1 \alpha^{n-1}.$$

So $\alpha > a_1 \geq 1$. Now we show that all other zeros β of $f(x)$ satisfy $|\beta| < 1$.

Consider the polynomial $g(x) = (x-1)f(x)$. Clearly,

$$g(x) = x^{n+1} - b_1 x^n + b_2 x^{n-1} + \cdots + b_{n+1},$$

where $b_1 = a_1 + 1, b_2 = a_1 - a_2, \cdots, b_n = a_{n-1} - a_n, b_{n+1} = a_n$. The numbers $b_1, \cdots, b_{n+1}$ are non-negative integers and $b_1 = 1 + b_2 + \cdots + b_{n+1}$ with $b_{n+1} \geq 1$. Let

$$h(z) = b_1 z^n - b_2 z^{n-1} - \cdots - b_{n+1}.$$

For all sufficiently small $\epsilon > 0$, first we show that

$$|h(z)| > |z^{n+1}| = |g(z) + h(z)|, \text{ if } |z| = 1 + \epsilon.$$

Indeed, if $|z| = 1 + \epsilon$, then

$$|h(z)| - |z^{n+1}| \geq b_1(1+\epsilon)^n - b_2(1+\epsilon)^{n-1} - \cdots - b_{n+1} - (1+\epsilon)^{n+1}$$

$$= \epsilon(nb_1 - (n-1)b_2 - \cdots - 2b_{n-1} - b_n - (n+1))$$

$$+ \text{ (higher terms in } \epsilon)$$

$$= \epsilon(b_2 + 2b_3 + \cdots + (n-1)b_n + nb_{n+1} - 1)$$

$$+ \text{ (higher terms in } \epsilon).$$

The coefficient of ϵ is positive. For sufficiently small $\epsilon > 0$, we have $|h(z)| - |z^{n+1}| > 0$. In this case

$$|g(z) + h(z)| = |z^{n+1}| < |h(z)| \leq |g(z)| + |h(z)|.$$

Therefore, by Rouche's theorem, the polynomial $g(z)$ and $h(z)$ have the same number of roots inside the disk $|z| \leq 1 + \epsilon$.

If $|z| \geq 1$, then

$$|h(z)| \geq b_1|z|^n - b_2|z|^{n-1} - \cdots - b_{n+1}$$

$$\geq |z|^n(b_1 - b_2 - \cdots - b_{n+1}) = |z|^n > 0.$$

So all the n roots of $h(z)$ lie strictly inside the unit disk $|z| \leq 1$. Letting $\epsilon \to 0$, we see that inside and on the boundary of the unit disk there are exactly n roots of the polynomial $g(x) = (x-1)f(x)$.

If $|z| = 1$ and $z \neq 1$, we claim that $g(z) \neq 0$. Otherwise there is z_0 such that $|z_0| = 1$, $z_0 \neq 1$ and $g(z_0) = 0$. Then

$$|z_0^{n+1} - b_1 z_0^n| = |b_2 z_0^{n-1} + \cdots + b_n z_0 + b_{n+1}|,$$

$$|z_0^{n+1} - b_1 z_0^n| > b_1 - 1 = b_2 + \cdots + b_{n+1}$$

$$\geq |b_2 z_0^{n-1} + \cdots + b_n z_0 + b_{n+1}|,$$

which is a contradiction. Hence exactly $n - 1$ roots of $f(x)$ lie inside the unit disk.

Using the similar arguments in the proof of Theorem 1.8.1, we see that $f(x)$ is irreducible. □

We remark that the proofs for Lemma 1.8.3 and Theorem 1.8.7 are modified from that in [P].

Example 1.8.2. *For any positive integer n, show that $x^{2n} - 2x^n - 7$ is irreducible in $\mathbb{Z}[x]$.*

Proof. The constant term 7 is a prime satisfying $7 > 1 + |-2|$. From Osada's Irreducibility Criterion we know that $x^{2n} - 2x^n - 7$ is irreducible in $\mathbb{Z}[x]$. □

1.9. Exercises.

(1) Let R be a ring that contains at least two elements. Suppose for each $a \in R$ there is a unique $\varphi(a) \in R$ such that $a\varphi(a)a = a$. Show that R is a division ring.

Hints: You may follow the steps below.

(a). Show that R has no zero divisors.

(b). Show that $\varphi(a)a\varphi(a) = \varphi(a)$.

(c). Show that R has unity.

(d). Show that R is a division ring.

(2) Let R be a unital ring. If $(xy)^2 = x^2y^2$ for all $x, y \in R$, show that R is commutative.

(3) Let R be a unital ring such that $x^6 = x$ for all $x \in R$.

(a). Prove that $x^2 = x$ for all $x \in R$.

(b). Prove that R is commutative.

(4) Define the center of a ring R as $Z(R) = \{a \in R : ab = ba, \forall b \in R\}$. Prove that a ring R is commutative if $a^2 - a \in Z(R)$, for all $a \in R$.

(5) Let $(R, +, \cdot)$ be a finite ring with at least two distinct elements. Suppose that the multiplication satisfies the cancellation rules. Show that R is a division ring.

(6) Consider $(S, +, \cdot)$, where S is a set and $+$ and $\cdot$ are binary operations on S such that

(a). $(S, +)$ is a group,

(b). $(S^*, \cdot)$ is a group where $S^* = S \backslash \{0\}$ and 0 is the additive identity element,

(c). $a(b + c) = (ab) + (ac)$ and $(a + b)c = (ac) + (bc)$ for all $a, b, c \in S$.
Show that $(S, +, \cdot)$ is a division ring.

(7) Show that in any unital ring R the commutative law for addition is redundant, in the sense that it follows from the other axioms for a ring.

(8) Determine all irreducible polynomials of degree 3 in $\mathbb{Z}_2[x]$. Justify your answer.

(9) Let R be a finite ring without zero-divisors. Show that $2\sum_{r \in R} r = 0$.

(10) Let a, b be elements of the unital ring R. Show that $1 - ab$ is a unit if and only if $1 - ba$ is a unit.

(11) Find all real roots of the polynomial $2x^4 + 3x^3 + 3x - 2 \in \mathbb{R}[x]$.

(12) Factor the polynomial $f(x) = x^5 + 3x^3 + x^2 + 2x \in \mathbb{Z}_5[x]$ into a product of irreducible polynomials.

(13) Whether is the rational polynomial $x^3 + 3x^2 - 8$ irreducible?

(14) Let R be a commutative ring, let a be a unit of R, and let b be any element of R. Define a function $\phi : R[x] \to R[x]$ by $\phi(f(x)) = f(ax + b)$, for all $f(x) \in R[x]$. Show that ϕ is an automorphism of $R[x]$.

(15) In the polynomial ring $\mathbb{Z}[x]$, show that the ideal $\langle n, x \rangle$ generated by $n \in \mathbb{Z}$ and x is a prime ideal if and only if n is a prime number.

(16) Show that any principal ideal in the polynomial ring $\mathbb{Z}[x]$ cannot be a maximal ideal.

(17) Let P be a prime ideal of a commutative ring R. Prove that $A \cap B \subseteq P$ implies $A \subseteq P$ or $B \subseteq P$, for all ideals A, B of R. Give an example to show that the converse is false.

(18) Let R be a commutative unital ring. Prove that if every proper ideal of R is prime, then R is a field.

(19) Let R be a commutative ring with 1. Show that the intersection of all the prime ideals of R is precisely the set of nilpotent elements of R.

(20) Let R be a commutative ring with 1 and let $\{P_i\}$ be a chain of prime ideals in R. Prove that $\cap P_i$ is prime. Deduce that every prime ideal P of R contains a minimal prime ideal of R.

(21) Let R be a commutative ring with 1 and let I be the intersection of all the maximal ideals of R. Prove that $a \in I$ if and only if $1 + ax$ is a unit in R for all $x \in R$.

(22) Let R be a finite unital ring, and $x \in R$ be nonzero. Show that x is a left 0-divisor if and only if it is a right 0-divisor.

(23) Let $\mathbb{R}$ denote the real numbers. What familiar ring is isomorphic to $\mathbb{R}[x]/\langle x^2 - x + 1 \rangle$? Prove your assertion. Prove that $\mathbb{Z}[x]/\langle x^2 + 1 \rangle$ is an integral domain, and find its cardinality.

(24) Let $\mathbb{Q}^n$ be the space of all n-tuples of rational numbers, made into a ring by component-wise addition and multiplication. Find all ring homomorphisms of $\mathbb{Q}^n$ onto $\mathbb{Q}$ and all ring homomorphisms of $\mathbb{Q}^n$ onto $\mathbb{Q}^n$.

(25) Show that the ideal of $\mathbb{Z}[x]$ generated by 2 and $x^4 + x^2 + 1$ is not maximal.

(26) Let $R_1 = \mathbb{Z}_p[x]/\langle x^2 - 3 \rangle, R_2 = \mathbb{Z}_p[x]/\langle x^2 + 2 \rangle$. Determine whether R_1 and R_2 are isomorphic for $p = 2, 5, 11$ respectively.

(27) Let F be one of the following fields $\mathbb{R}, \mathbb{Q}, \mathbb{C}, \mathbb{Z}_9$. Let $I \subset F[x]$ be the ideal generated by $x^4 + 2x - 2$. For which choices of F is the ring $F[x]/I$ a field?

(28) Let $R = \mathbb{C}([0,1])$ be the ring of continuous real-valued functions on the interval $[0,1]$ with the usual definition of sum and product of functions from calculus. Show that $f \in R$ is a zero divisor if and only if f is not identically zero and $\{x : f(x) = 0\}$ contains an open interval. What are the idempotents of this ring? What are the nilpotents? What are the units?

(29) Let $R = (\mathbb{Z}_2, +, \cdot)$ and let $G = (\mathbb{Z}_2, +)$, the group of order 2. Find a quotient of a polynomial ring which is isomorphic to the group ring $R[G]$.

(30) Let G be a finite elementary 2-group (i.e., a direct product of finitely many copies of $\mathbb{Z}_2$, the cyclic group of order 2).

(a). Show that $\mathbb{Z}[G]$ has zero divisors.

(b). Show that there is a one-to-one correspondence between ring homomorphisms $\psi : \mathbb{Z}[G] \to \mathbb{Z}$ and group homomorphisms $\chi : G \to \{\pm 1\}$.

(c). The augmentation mapping is the homomorphism $\mathbb{Z}[G] \to \mathbb{Z}$ defined by sending all $g \in G$ to 1. Let I be the kernel of this homomorphism. What are all the maximal ideals containing I?

(d). Let P be a minimal prime ideal of $\mathbb{Z}[G]$. You may assume that $P \cap \mathbb{Z} = 0$.

(e). Let M be the maximal ideal from (d) that contains I and 2. Show that $P \subset M$.

(31) Is $x^4 + 1$ irreducible over the field of real numbers? over the field of rational numbers? over a field with 16 elements?

(32) In $\mathbb{R}[x]$, consider the set of polynomials $f(x)$ for which $f(2) = f'(2) = f''(2) = 0$. Prove that this set forms an ideal and find its monic generator. Do the polynomials such that $f(2) = f'(3) = 0$ form an ideal? Prove or give a counterexample.

(33) Let R be a commutative ring with identity, and let A, B be ideals of R. The radical of A is defined to be $\sqrt{A} = \{r \in R : r^n \in A \text{ for some } n \geq 0\}$.
   (a). Show A is an ideal of R.
   (b). Show that if $\sqrt{A} + \sqrt{B} = R$, then $A + B = R$.

(34) Let R be a ring. Show that the two-sided ideals of the ring $M_n(R)$ of $n \times n$ matrices over R are precisely the subsets of the form

$$M_n(I) = \{(a_{ij}) : a_{ij} \in I \ \forall \ i, j\},$$

where I is some ideal of R.

(35) Show that the ring of even integers contains a maximal ideal M such that E/M is not a field.

(36) If N is the ideal of all nilpotent elements in a commutative ring R, show that R/N is a ring with no nonzero nilpotent elements.

(37) Show that the maximal ideals of the polynomial ring $\mathbb{C}[x]$ are principal ideals generated by $x - c$ for some $c \in \mathbb{C}$.

(38) Show that the prime ideals of $\mathbb{C}[x, y]$ are $\langle 0 \rangle, \langle f(x, y) \rangle$ for an irreducible $f(x, y) \in \mathbb{C}[x, y]$ and $\langle x - a, y - b \rangle$ for some $a, b \in \mathbb{C}$.

(39) Let R be a commutative ring. A **minimal prime ideal** in R is a prime ideal I such that if $J \subset I$ is a prime ideal of R then $I = J$. Show that every prime ideal I contains a nonzero minimal prime ideal. (Hints: Use Kuratowski-Zorn lemma.)

(40) Let $P_1, P_2, \cdots, P_n, n \geq 2$, be ideals in a ring R, with $P_3, \cdots, P_n$ prime (if $n \geq 3$). Let P be any ideal of R. If $P \subseteq \cup_{i=1}^n P_i$, show that $P \subseteq P_i$ for some i.

(41) Show that all nonzero prime ideals are maximal in a principal ideal domain.

(42) Let D be an integral domain for which $IJ = I \cap J$ for all ideals I, J of D. Prove that D is a field.

(43) A **valuation ring** R is a commutative unital ring in which, for any a and b in R, either $a|b$ or $b|a$. Show that the non-units (i.e., non-invertible elements) of a valuation ring R form an ideal that is necessarily the unique maximal ideal in R.

(44) Show that End $((\mathbb{Z}, +)) \cong (\mathbb{Z}, +, \cdot)$ and that $\text{End}((\mathbb{Z}_n, +)) \cong (\mathbb{Z}_n, +, \cdot)$.

(45) Show that $\text{End}((\mathbb{Z}_2 \times \mathbb{Z}_2, +))$ is not isomorphic to $(\mathbb{Z}_2 \times \mathbb{Z}_2, +, \cdot)$.

(46) Find a subring of the ring $(\mathbb{Z} \times \mathbb{Z}, +, \cdot)$ that is not an ideal of $\mathbb{Z} \times \mathbb{Z}$.

(47) Find all ideals of the ring $(F[x] \times F[x], +, \cdot)$ where F is a field.

(48) Let F be a field, and let S be any subset of $F \times F \times \cdots \times F$ for n factors. Show that the set N_S of all $f(x_1, \cdots, x_n) \in F[x_1, \cdots, x_n]$ that have every element $(a_1, \cdots, a_n)$ of S as a zero is an ideal in $F[x_1, \cdots, x_n]$.

(49) How many maximal ideals of $\mathbb{Z}[x]$ contain $\{42, x^2 + 1\}$?

(50) Is $\mathbb{Q}[x]/\langle x^3 - 6x - 6 \rangle$ a field? Why?

(51) Find all $c \in \mathbb{Z}_7$ so that $\mathbb{Z}_7[x]/\langle x^2 + x + c \rangle$ is a field.

(52) For any $n \in \mathbb{Z}^+$, show that $x^{2^n} + 1$ is irreducible over $\mathbb{Q}$.

(53) Let be p a prime. For any monic $f(x) \in \mathbb{Z}[x]$ with $\deg(f(x)) = p - 1$, show that $\int_0^x f(t)dt + 1$ is irreducible over $\mathbb{Q}$.

(54) Is $x^4 + x^3 + x^2 + x + 1$ is irreducible in $\mathbb{Q}[x]$? in $\mathbb{Z}_2[x]$?

(55) Factor the polynomial $f(x) = x^{12} + x^9 + x^6 + x^3 + 1$ into product of irreducibles over in $\mathbb{Q}[x]$. (Hint: Consider $(x^3 - 1)f(x)$.)

(56) Factor the polynomial $x^7 + 1$ into product of irreducibles over in $\mathbb{Z}_2[x]$.

(57) Is $x^4 + x + 1$ is irreducible in $\mathbb{Q}[x]$? in $\mathbb{Z}_2[x]$?

(58) Show that $f(x) = (1 + x + \cdots + x^n)^2 - x^n \in \mathbb{Z}[x]$ is a product of two polynomials in $\mathbb{Z}[x]$. (Hints: Consider $(x - 1)^2 f(x)$.)

(59) Let $n \geq 2$ be an integer. Show that the polynomial $f_n(x) = x^{n-1} + x^{n-2} + \cdots + x + 1 \in \mathbb{Q}[x]$ is irreducible if and only if n is a prime number.

(60) For any integer $n > 2$, show that all roots of $f(x) = x^{n-1} + 2x^{n-2} + 3x^{n-3} + \cdots + (n-1)x + n$ have norm larger than 1. (Hints: Consider $(x - 1)f(x)$.)

(61) For any prime p, show that
$$f(x) = x^{p-1} + 2x^{p-2} + 3x^{p-3} + \cdots + (p-1)x + p$$
is irreducible in $\mathbb{Z}[x]$. (Hints: Use the previous question.)

(62) Let a be a nonzero integer, and $n \geq 3$ be another integer. Show that the polynomial $f(x) = x^n + ax^{n-1} + ax^{n-2} + \cdots + ax - 1$ is irreducible over the integers. (Hints: Use Brauer's Theorem twice.)

(63) In Theorem 1.8.6, if we replace the condition "$p > |a_1| + \cdots + |a_n|$" with "$p = |a_1| + \cdots + |a_n|$ and no root of $f(x)$ is a root of unity", show that $f(x)$ is also irreducible over $\mathbb{Q}$.

2. Unique Factorization Domains

In this chapter we will study an important class of integral domains: unique factorization domains (UFD, for short). We further investigate some special UFDs: PID and Euclidean domains. We also provide some applications for them. We remark that many materials in this chapter and Chapter 4 have appeared in Chapters 5 and 6 in [LZ].

2.1. Basic definitions.

We already knew that the ring $\mathbb{Z}$ of integers and the polynomial ring $F[x]$ over a field F are integral domains in which every element can be factored into unique product of irreducibles. Now we consider the general case.

Recall that $\mathcal{U}(R)$ is the unit group of a unital ring R.

Definition 2.1.1. *Let R be a unital commutative ring. Two elements $a, b \in R$ are* **associates** *in R if $a = b\varepsilon$ for some $\varepsilon \in \mathcal{U}(R)$, denoted by $a \sim b$.*

It is easy to show that the relation $a \sim b$ is an equivalence relation on R.

Example 2.1.1. *We know that $\mathcal{U}(\mathbb{Z}) = \{\pm 1\}$. So the only associates of 6 in $\mathbb{Z}$ are ± 6.*

The following easy results will be repeatedly used later.

Lemma 2.1.2. *For nonzero elements a and b of an integral domain D, we have*

(i). $\langle a \rangle \subseteq \langle b \rangle$ if and only if $b | a$, and
(ii). $\langle a \rangle = \langle b \rangle$ (or equivalently, $a | b$ and $b | a$) if and only if $a \sim b$.

Proof. (i). Note that $\langle a \rangle \subseteq \langle b \rangle = bD$ if and only if $a \in \langle b \rangle$, if and only if $a = bd$ for some $d \in D$, if and only if $b | a$.

(ii). Using (a), we see that $\langle a \rangle = \langle b \rangle$ if and only if $a = bc$ and $b = ad$ for some $c, d \in D$. But then $a = adc$ and by canceling, we obtain $1 = dc$. Thus d and c are units so $a \sim b$. $\qquad\square$

Definition 2.1.3. *Let D be an integral domain and $0 \neq a \in D$. All units and associates of a are called the* **trivial factors** *of a.*

Let D be an integral domain. Recall from Definition 1.1.20 that an element $a \in D$ is irreducible if $a = bc$ for some $b, c \in D$ implies that b or c is a unit.

Lemma 2.1.4. *Let D be an integral domain. If $p \in D$ is irreducible and $p \sim q$, then q is irreducible.*

Proof. Since $p \sim q$, then $p = \varepsilon q$ for an $\varepsilon \in \mathcal{U}(D)$. Any factorization $q = ab$, where $a, b \in D$, gives $p = \varepsilon ab$. Clearly a and b cannot be both units. We may assume that b is not. We deduce that εa is a unit since p is irreducible, and further a is a unit. Thus q is also an irreducible. $\qquad\square$

Definition 2.1.5. *Let D be an integral domain. We say that D is a* **unique factorization domain** *(UFD for short) if the following conditions are satisfied.*

(i). Every element of D that is neither 0 nor a unit can be factored into a product of a finite number of irreducibles.

(ii). If $p_1 \cdots p_s = q_1 \cdots q_t$ where p_i, q_j are irreducibles in D, then $s = t$ and the q_j can be renumbered so that $p_i \sim q_i$.

Example 2.1.2. *If F is a field, we know that the ring $F[x]$ is a UFD. Also we know that the integer $\mathbb{Z}$ is a UFD. Consider in $\mathbb{Z}$ the factorizations*

$$30 = (2)(3)(5) = (-2)(-3)(5) = (3)(-2)(-5).$$

Clearly $2 \sim -2$, $3 \sim -3$ and $5 \sim -5$. Thus except for order and associates, the irreducible factors in these three factorizations of 30 are the same.

Notice that in a UFD D, any nonzero nonunit element $a \in D$ can be written as

$$a = \varepsilon p_1^{r_1} p_2^{r_2} \cdots p_s^{r_s},$$

where ε is a unit, $r_i \geq 1$, $p_1, p_2, \cdots, p_s$ are irreducibles and not associates.

Definition 2.1.6. *A nonzero nonunit element a of an integral domain D is called a* **prime** *if, for all $b, c \in D$, $a|bc$ implies either $a|b$ or $a|c$.*

Note that, $a \in D$ is prime if and only if $\langle a \rangle$ is a prime ideal of D.

Lemma 2.1.7. *Let D is an ID. If $p \in D$ is prime, then p is an irreducible.*

Proof. Suppose that $p = ab$ for some $a, b \in D$. Since p is prime, then $p|a$ or $p|b$. We may assume that $p|a$, i.e., $a = pc$ for some $c \in D$. We have $p = pbc$. So $bc = 1$, and $b \in \mathcal{U}(D)$. Thus p is an irreducible. $\square$

One can show that in a UFD an irreducible is also a prime (see Exercise 2.6(2)). Thus the concepts of prime and irreducible coincide in a UFD. The concepts do not coincide in every domain.

Example 2.1.3. *Consider the subdomain $D = \mathbb{R}[x^2, xy, y^2]$ of $\mathbb{R}[x, y]$. Then x^2, xy, and y^2 are irreducibles in D, but*

$$(x^2)(y^2) = (xy)(xy).$$

Since xy divides $x^2 y^2$ but not x^2 or y^2, we see that xy is not a prime in D. Similarly, neither x^2 nor y^2 is a prime.

Definition 2.1.8. *Let D be a UFD and let $a_1, a_2, \cdots, a_n$ be nonzero elements of D. An element d of D is called a **greatest common divisor** (abbreviated gcd) of all of the a_i if the following are satisfied.*

(i). $d|a_i$ for $i = 1, \cdots, n$,
(ii). $d' \in D$ and $d'|a_i$ imply that $d'|d$.

When we write $d \sim \gcd(a_1, a_2, \cdots, a_n)$ we mean that d is one of the gcds of $a_1, a_2, \cdots, a_n$.

We can easily see that any two gcd's are associates. The well-known technique in the example below shows that gcd's exist in a UFD.

Example 2.1.4. *We knew how to find $\gcd(420, -168, 252)$ in $\mathbb{Z}$. Factoring, we obtain $420 = 2^2 \cdot 3 \cdot 5 \cdot 7, -168 = 2^3 \cdot (-3) \cdot 7$, and $252 = 2^2 \cdot 3^2 \cdot 7$. Then $\gcd(420, -168, 252) = 4 \cdot 3 \cdot 1 \cdot 7 = 84$. The only other gcd of these numbers in $\mathbb{Z}$ is -84, because 1 and -1 are the only units.*

The technique in the above Example depends on being able to factor an element of a UFD into a product of irreducibles. This can be a tough job, even in $\mathbb{Z}$. Later we will learn a technique, the Euclidean Algorithm, that will allow us to find gcd's in some UFDs.

2.2. Principal ideal domains.

Let R be a commutative unital ring. Let $a \in R$. Recall that the principal ideal $\langle a \rangle$ consists of all multiples of the element a, i.e., $\langle a \rangle = Ra$. From Definition 1.5.10 we know that an integral domain D is a principal ideal domain (abbreviated PID) if every ideal in D is a principal ideal. We see that $\mathbb{Z}$ is a PID because every ideal is of the form $n\mathbb{Z}$, the ideal generated by some integer n. If F is a field, then $F[x]$ is a PID (Theorem 1.5.11). Our purpose in this section is to prove that every PID is a UFD.

Let us start with an easy result.

Lemma 2.2.1. *Let $I_1 \subseteq I_2 \subseteq \cdots$ be an ascending chain of ideals I_i in a ring R. Then $N = \cup_i I_i \trianglelefteq R$.*

Proof. Let $a, b \in N, r \in R$. Then there are ideals I_i and I_j in the chain, with $a \in I_i$ and $b \in I_j$. Now either $I_i \subseteq I_j$ or $I_j \subseteq I_i$. We may assume that $I_i \subseteq I_j$, so both a and b are in I_j. This implies that $a \pm b, ar, ra \in I_j$, so $a \pm b, ar, ra \in N$. Hence $N \trianglelefteq R$.   □

Lemma 2.2.2 (Ascending Chain Condition). *Let D be a PID, and $I_1 \subseteq I_2 \subseteq \cdots$ be an ascending chain of ideals I_i of D. Then there exists a positive integer m such that $I_n = I_m$ for all $n \geq m$.*

Proof. By the above Lemma, we know that $N = \cup_i I_i \trianglelefteq D$. Since D is a PID, there is a $c \in D$ such that $N = \langle c \rangle$. Since $N = \cup_i I_i$, we must have $c \in I_m$, for some $m \in \mathbb{N}$. For $n \geq m$, we have

$$\langle c \rangle \subseteq I_m \subseteq I_n \subseteq N = \langle c \rangle.$$

Thus $I_m = I_n$ for $n \geq m$.   □

The above lemma tells us that every strictly ascending chain of ideals (that is, all inclusions proper) in a PID is of finite length. In this situation we say that the **ascending chain condition (ACC)** holds for ideals in the PID.

Theorem 2.2.3. *Let D be a PID, and $a \in D$ neither 0 nor a unit. Then a can be factored into a product of irreducibles.*

Proof. Claim 1. a has at least one irreducible factor.

If a is an irreducible, we are done. If a is not an irreducible, then $a = a_1 b_1$, for some nonunit elements $a_1, b_1 \in D$. Now $\langle a \rangle \subset \langle a_1 \rangle$. In this manner, then starting now with a_1, we arrive at a strictly ascending chain of ideals

$$\langle a \rangle \subset \langle a_1 \rangle \subset \langle a_2 \rangle \subset \cdots .$$

By Lemma 2.2.2, this chain terminates with some $\langle a_r \rangle$, and a_r must then be irreducible. Thus a has an irreducible factor a_r. Claim 1 follows.

From Claim 1, either a is irreducible or $a = p_1 c_1$ for p_1 an irreducible and c_1 not a unit. We see that $\langle a \rangle \subset \langle c_1 \rangle$. If c_1 is not irreducible, then $c_1 = p_2 c_2$ for an irreducible p_2 with c_2 not a unit. Continuing in this manner, we get a strictly ascending chain of ideals

$$\langle a \rangle \subset \langle c_1 \rangle \subset \langle c_2 \rangle \subset \cdots .$$

By Lemma 2.2.2 this chain must terminate, i.e., with some $c_r = q_r$ that is an irreducible. Thus $a = p_1 p_2 \cdots p_r q_r$. $\qquad \square$

We first give the following result which is of some interest in itself.

Lemma 2.2.4. *Let D is a PID and $p \in D \setminus \{0\}$. Then $\langle p \rangle$ is maximal in D if and only if p is an irreducible.*

Proof. ($\Rightarrow$). Let $\langle p \rangle$ be a maximal ideal of D. Suppose that $p = ab$ where $a, b \in D$. It follows that $\langle p \rangle \subset \langle a \rangle$. Suppose that $\langle a \rangle = \langle p \rangle$. Then $a \sim p$, so b must be a unit. If $\langle a \rangle \neq \langle p \rangle$, we see that $\langle p \rangle \subset \langle a \rangle$, i.e., $\langle a \rangle = \langle 1 \rangle = D$, since $\langle p \rangle$ is maximal. Then a is a unit. Thus, p is an irreducible of D.

($\Leftarrow$). Suppose that p is an irreducible in D. If $\langle p \rangle \subseteq \langle a \rangle \trianglelefteq D$ for some $a \in D$, we have $p = ab$ for some $b \in D$. Since p is an irreducible, we see that a or b is a unit.

Now if a is a unit, then $\langle a \rangle = \langle 1 \rangle = D$.

If b is a unit, then $a = b^{-1} p$, so $\langle a \rangle \subseteq \langle p \rangle$, and we have $\langle a \rangle = \langle p \rangle$. Thus $\langle p \rangle \subseteq \langle a \rangle$ implies that either $\langle a \rangle = D$ or $\langle a \rangle = \langle p \rangle$. Hence $\langle p \rangle$ is a maximal ideal. $\qquad \square$

Lemma 2.2.5. *Let D be a PID and $p \in D$. If p is an irreducible, then p is prime.*

Proof. Suppose that $p | ab$ where $a, b \in D$. Then $ab \in \langle p \rangle$. Since p is an irreducible, we know that $\langle p \rangle$ is a maximal ideal. From Theorems 1.7.2 and 1.5.4, then $\langle p \rangle$ is a prime ideal. So either $a \in \langle p \rangle$ or $b \in \langle p \rangle$, yielding that either $p | a$ or $p | b$. Thus, p is prime. $\qquad \square$

Corollary 2.2.6. *If p is an irreducible in a PID D and $p | a_1 a_2 \cdots a_n$ for $a_i \in D$, then $p | a_i$ for some i.*

Proof. This is immediate from using mathematical induction. $\square$ Now we can prove the main result in this section.

Theorem 2.2.7. *Every PID is a UFD.*

Proof. Let D be a PID and $a \in D$, where a is neither 0 nor a unit. Then a has a factorization $a = p_1 p_2 \cdots p_r$ into irreducibles. We need only to show uniqueness. Let

$$p_1 p_2 \cdots p_r = q_1 q_2 \cdots q_s$$

where p_i, q_j are irreducibles. Then we have $p_1 | q_1 q_2 \cdots q_s$, which implies that $p_1 | q_j$ for some j. By changing the order of the q_j if necessary, we can assume that $j = 1$ so $p_1 | q_1$. Then $q_1 = p_1 u_1$, and since q_1, p_1 are irreducibles, u_1 is a unit, so $p_1 \sim q_1$. We have

$$p_1 p_2 \cdots p_r = p_1 u_1 q_2 \cdots q_s.$$

Then

$$p_2 \cdots p_r = u_1 q_2 \cdots q_s.$$

In this manner, we will arrive at $p_i \sim q_i$ for $i \leq r$ and

$$1 = u_1 u_2 \cdots u_r q_{r+1} \cdots q_s.$$

Since each q_j is irreducible, we must have $r = s$. $\square$

The converse to the above theorem is false. That is, a UFD need not be a PID (see Example 2.4.3).

The following well-known result from elementary number theory can easily follow from the fact that $\mathbb{Z}$ is a PID.

Corollary 2.2.8 (Fundamental Theorem of Arithmetic). *The ring $\mathbb{Z}$ of integers is a UFD.*

2.3. Euclidean domains.

In this section we shall study an important class of UFDs, the Euclidean domains.

Definition 2.3.1. *Let D be an ID. A **Euclidean norm** on D is a function $\nu : D^* \to \mathbb{Z}^+$ such that the following conditions are satisfied for all $a, b \in D^* = D \setminus \{0\}$.*

 (i). There exist $q, r \in D$ such that $a = bq + r$, where either $r = 0$ or $\nu(r) < \nu(b)$.
 (ii). $\nu(a) \leq \nu(ab)$.

*An integral domain D is a **Euclidean domain** if it has a Euclidean norm.*

Example 2.3.1. *The integer domain* $\mathbb{Z}$ *is a Euclidean domain, since it has a Euclidean norm*

$$\nu : \mathbb{Z} \to \mathbb{Z}^+, \nu(n) = |n|, \forall n \in \mathbb{Z}.$$

Condition 1 holds by the division algorithm for $\mathbb{Z}$*. Condition (ii) follows from* $|ab| = |a| \cdot |b|$ *and* $|b| \geq 1$ *for* $b \neq 0$ *in* $\mathbb{Z}$*.*

Example 2.3.2. *For any field* F*, the polynomial ring* $F[x]$ *is a Euclidean domain, since from Division Algorithm* $F[x]$ *has a Euclidean norm*

$$\nu : F[x]^* \to \mathbb{Z}^+, \nu(f(x)) = \deg(f(x)), \forall f(x) \in F[x]^*.$$

Of course, we shall have some examples of Euclidean domains other than these familiar ones later.

Theorem 2.3.2. *Every Euclidean domain* D *is a PID.*

Proof. Let D have a Euclidean norm ν, and let $I \trianglelefteq D$. If $I = 0$, then $I = \langle 0 \rangle$. Now suppose that $I \neq 0$. Take $b \in I \setminus \{0\}$ such that $\nu(b)$ is minimal among all $\nu(g)$ for $g \in I \setminus \{0\}$. We claim that $I = \langle b \rangle$. Let $a \in I$. Then by Condition (i) for a Euclidean domain, there exist $q, r \in D$ such that $a = bq + r$, where either $r = 0$ or $\nu(r) < \nu(b)$. Now $r = a - bq$ and $a, b \in I$, so that $r \in I$ since $I \trianglelefteq D$. So $\nu(r) < \nu(b)$ is impossible by the choice of b. Thus $r = 0$, so $a = bq$, and $a \in \langle b \rangle$. Hence, $I \subseteq \langle b \rangle \subseteq I$, and further $I = \langle b \rangle$. $\square$

Corollary 2.3.3. *A Euclidean domain is a UFD.*

Finally, we should mention that examples of PIDs that are not Euclidean are not easily found, however.

Let D be a Euclidean domain with a Euclidean norm ν. We can use Condition (ii) of a Euclidean norm to characterize the units of D.

Theorem 2.3.4. *Let* D *be a Euclidean domain with a Euclidean norm* ν*.*

(i). $\nu(1) \leq \nu(a)$ *for any nonzero* $a \in D$*.*
(ii). $u \in D$ *is a unit if and only if* $\nu(u) = \nu(1)$*.*

Proof. (i). For $a \in D \setminus \{0\}$, we have $\nu(1) \leq \nu(1a) = \nu(a)$.
(ii). ($\Rightarrow$). If $u \in \mathcal{U}(D)$, then

$$\nu(u) \leq \nu(uu^{-1}) = \nu(1).$$

Thus $\nu(u) = \nu(1)$.

($\Leftarrow$). Suppose $u \in D$ with $\nu(u) = \nu(1)$. Then by the division algorithm, there exist $q, r \in D$ such that

$$1 = uq + r,$$

where either $r = 0$ or $\nu(r) < \nu(u)$. By (i), we know that $\nu(r) < \nu(u)$ is impossible. Then $r = 0$ and $1 = uq$. Hence $u \in \mathcal{U}(D)$. $\square$

Example 2.3.3.   (a). For $\mathbb{Z}$ with $\nu(n) = |n|$, the minimum of $\nu(n)$ for nonzero $n \in \mathbb{Z}$ is 1. Thus, ± 1 are the only elements of $\mathbb{Z}$ with $\nu(n) = 1$. So ± 1 are the only units of $\mathbb{Z}$ by Theorem 2.3.4.

   (b). The polynomial ring $F[x]$ with $\nu(f(x)) = \deg(f(x))$ for $f(x) \neq 0$ is a Euclidean domain. The minimum value of $\nu(f(x))$ for all nonzero $f(x) \in F[x]$ is 0. The nonzero polynomials of degree 0 are exactly the nonzero elements of F, So $F^* = \mathcal{U}(F[x])$ by Theorem 2.3.4.

We know that in any UFD D, $\gcd(a, b)$ exists for any $a, b \in D$. But it is generally very hard to find $\gcd(a, b)$. The best property for a Euclidean domain is that there is a nice algorithm for this, as the next theorem shows.

Theorem 2.3.5 (Euclidean Algorithm). *Let D be a Euclidean domain with a Euclidean norm ν, and let a and b be nonzero elements of D.*

   (i). There are $q_i, r_i \in D$ such that

$$a = bq_1 + r_1,$$
$$b = r_1 q_2 + r_2,$$
$$r_1 = r_2 q_3 + r_3,$$
$$\cdots \tag{2.1}$$
$$r_{s-3} = r_{s-2} q_{s-1} + r_{s-1},$$
$$r_{s-2} = r_{s-1} q_s + r_s,$$

where $r_s = 0$, $\nu(r_{s-1}) < \nu(r_{s-2}) < \cdots < \nu(r_2) < \nu(r_1) < \nu(b)$. Furthermore $\gcd(a, b) \sim r_{s-1}$.

   (ii). If $\gcd(a, b) \sim d$, then there exist $\lambda, \mu \in D$ such that $d = \lambda a + \mu b$.

Proof.   (i).   Since $\nu(r_i) < \nu(r_{i-1})$ and $\nu(r_i)$ is a nonnegative integer, it follows that after some finite number of steps we must arrive at some $r_s = 0$. Thus, we have all equations in (2.1).

Suppose $d \sim \gcd(a, b)$. From $d|a$ and $d|b$, we have $d|r_1$. From $d|b$ and $d|r_1$, we have $d|r_2$. In this manner we deduce that $d|r_i$ for any i. In particularly $d|r_{s-1}$.

On the other hand $r_{s-1}|r_{s-2}$. From (2.1) backward we deduce that

$$r_{s-1}|r_{s-2}, r_{s-1}|r_{s-3}, \cdots, r_{s-1}|b$$

and $r_{s-1}|a$. Thus $r_{s-1}|d$. Therefore $r_{s-1} \sim d$.

(ii). We may assume that $d = r_{s-1}$. We shall prove by induction on k that $r_k = \lambda_k a + \mu_k b$ for some $\lambda_k, \mu_k \in D$. If $s = 1$, i.e., $r_1 = 0$, then $d = b$, and $d = 0a + 1b$ and we are done. Suppose that $r_j = \lambda_j a + \mu_j b$ for $j = 1, 2, \cdots, k$. Use $r_{k-1} = r_k q_{k+1} + r_{k+1}$ we deduce that

$$r_{k+1} = r_{k-1} - r_k q_{k+1} = (\lambda_{k-1} a + \mu_{k-1} b) - q_{k+1}(\lambda_k a + \mu_k b)$$
$$= \lambda_{k+1} a + \mu_{k+1} b.$$

Thus

$$d = r_{s-1} = \lambda_{s-1} a + \mu_{s-1} b$$

where $\lambda_{s-1}, \mu_{s-1} \in D$. □

Example 2.3.4. *Use Euclidean Algorithm in $\mathbb{Q}[x]$ to find a $\gcd(f(x), g(x))$, where*

$$f(x) = x^4 + x^3 - x^2 - 1, \; g(x) = x^3 + x^2 - 2x.$$

Solution. Notice that

$$x^4 + x^3 - x^2 - 1 = x(x^3 + x^2 - 2x) + (x - 1),$$
$$x^3 + x^2 - 2x = (x^2 + 2x)(x - 1) + 0.$$

So $\gcd(f(x), g(x)) = x - 1$. □

2.4. Polynomial rings over UFDs.

In this section we shall prove that polynomial rings over UFDs are UFDs. We always assume that D is a UFD.

Definition 2.4.1. *Let D be a UFD, and let*

$$f(x) = a_0 + a_1 x + \cdots + a_n x^n \in D[x] \setminus D.$$

*An element $c \in D$ is a **content** of $f(x)$ if $c \sim \gcd(a_0, a_1, \cdots, a_n)$. We say that $f(x)$ is **primitive** if $\gcd(a_0, a_1, \cdots, a_n) \sim 1$.*

Example 2.4.1. *In $\mathbb{Z}[x]$, $x^2 + 3x + 2$ is primitive, but $4x^2 + 2x + 8$ is not, since 2, a nonunit in $\mathbb{Z}$, is a common divisor of the coefficients 4, 2, and 8.*

Observe that every irreducible in $D[x]$ of positive degree must be a primitive polynomial.

Lemma 2.4.2. *Let D be a UFD. Then for every nonconstant $f(x) \in D[x]$ we have $f(x) = cg(x)$, where $c \in D, g(x) \in D[x]$, and $g(x)$ is primitive. Also $g(x)$ is unique up to a unit factor in D.*

Proof. Let $f(x) = a_0 + a_1 x + \cdots + a_n x^n \in D[x]$, where $a_0, a_1, \ldots, a_n$ with $a_n \neq 0$ and $n \geq 1$. Let $c \sim \gcd(a_0, a_1, \ldots, a_n)$. Write $a_i = cq_i$ for some $q_i \in D$. We have $f(x) = cg(x)$, where no irreducible in D divides all of the coefficients $q_0, q_1, \ldots, q_n$ of $g(x)$. So $g(x)$ is a primitive polynomial.

For uniqueness, if $f(x) = cg(x) = dh(x)$ for $c, d \in D, h(x), g(x) \in D[x]$, and $g(x), h(x)$ primitive. Since both c and d are contents of $f(x)$, then $d = cu$ for a unit $u \in D$. We see that $g(x) = uh(x)$ for a unit $u \in D$. From $f(x) = cg(x)$, we see that the primitive polynomial $g(x)$ is also unique up to a unit factor. $\square$

Example 2.4.2. *In $\mathbb{Z}[x]$, $4x^2 + 6x - 2 = 2(2x^2 + 3x - 1)$, where $2x^2 + 3x - 1$ is primitive.*

Lemma 2.4.3 (Gauss's Lemma). *Let D be a UFD, and $f(x), g(x)$ be two primitive polynomials in $D[x]$. Then $f(x)g(x)$ is also primitive.*

Proof. Let

$$f(x) = a_0 + a_1 x + \cdots + a_n x^n,$$
$$g(x) = b_0 + b_1 x + \cdots + b_m x^m,$$

and let $h(x) = f(x)g(x)$. Let p be an arbitrary irreducible in D. Then p does not divide all a_i, and p does not divide all b_j, since $f(x)$ and $g(x)$ are primitive. Let a_r be the first coefficient of $f(x)$ not divisible by p; i.e., $p|a_i$ for $i < r$, and $p \nmid a_r$. Similarly, let $p|b_j$ for $j < s$, and $p \nmid b_s$. The coefficient of x^{r+s} in $h(x) = f(x)g(x)$ is

$$c_{r+s} = (a_0 b_{r+s} + \cdots + a_{r-1} b_{s+1}) + a_r b_s + (a_{r+1} b_{s-1} + \cdots + a_{r+s} b_0).$$

Since $p|a_i$ for $i < r$, then

$$p|(a_0 b_{r+s} + \cdots + a_{r-1} b_{s+1}).$$

Since $p|b_j$ for $j < s$, then

$$p|(a_{r+1} b_{s-1} + \cdots + a_{r+s} b_0).$$

Since $p \nmid a_r$ or $p \nmid b_s$, so $p \nmid a_r b_s$, and consequently $p \nmid c_{r+s}$. This shows that any irreducible $p \in D$ does not divide some coefficient of $f(x)g(x)$. Therefore $f(x)g(x)$ is primitive. $\square$

Corollary 2.4.4. *Let D be a UFD. Then a finite product of primitive polynomials in $D[x]$ is also primitive.*

Proof. This follows from the above Lemma by induction.   □

Now let D be a UFD and let F be a field of quotients of D. Then we have known that $F[x]$ is a UFD.

Lemma 2.4.5. *Let D be a UFD and let F be a field of quotients of D. Let $f(x) \in D[x]$ with $\deg(f(x)) > 0$.*

(i). If $f(x)$ is an irreducible in $D[x]$, then $f(x)$ is also an irreducible in $F[x]$.

(ii). If $f(x)$ is primitive in $D[x]$ and irreducible in $F[x]$, then $f(x)$ is irreducible in $D[x]$.

Proof. (i). Suppose that $f(x) = r(x)s(x)$ for $r(x), s(x) \in F[x]$ with $\deg(r(x)) < \deg(f(x))$ and $\deg(s(x)) < \deg(f(x))$. Since F is a field of quotients of D, each coefficient in $r(x)$ and $s(x)$ is of the form a/b for some $a, b \in D$. By clearing denominators, we can get

$$df(x) = r_1(x)s_1(x)$$

for $d \in D$, and $r_1(x), s_1(x) \in D[x]$, where $\deg(r_1(x)) = \deg(r(x))$ and $\deg(s_1(x)) = \deg(s(x))$. Write $f(x) = ag(x), r_1(x) = a_1 r_2(x)$, and $s_1(x) = a_2 s_2(x)$ for primitive polynomials $g(x), r_2(x)$, and $s_2(x)$, and $a, a_1, a_2 \in D$. Then

$$(da)g(x) = a_1 a_2 r_2(x)s_2(x),$$

and $r_2(x)s_2(x)$ is primitive. By the uniqueness, $a_1 a_2 = dau$ for some unit u in D. So

$$(da)g(x) = daur_2(x)s_2(x),$$

yielding that

$$f(x) = ag(x) = aur_2(x)s_2(x).$$

This is impossible. Thus $f(x) \in D[x]$ is irreducible in $F[x]$.

(ii). A nonconstant $f(x) \in D[x]$ that is primitive in $D[x]$ and irreducible in $F[x]$ is also irreducible in $D[x]$, since $D[x] \subset F[x]$.   □

The above Lemma shows that if D is a UFD, the irreducibles in $D[x]$ are precisely the irreducibles in D, together with the nonconstant primitive polynomials that are irreducible in $F[x]$, where F is a field of quotients of D.

Corollary 2.4.6. *Let D be a UFD and let F be a field of quotients of D. Let $f(x) \in D[x]$ with $\deg(f(x)) > 0$. Then $f(x)$ factors into a*

product of two polynomials of degrees r and s in $F[x]$ if and only if
$f(x)$ has a factorization into polynomials of the same degrees r and
s in $D[x]$.

Proof. ($\Rightarrow$). This was shown in the proof for (i) of the previous
lemma.

($\Leftarrow$). This holds trivially since $D[x] \subseteq F[x]$.   □

Now we prove our main theorem in this section.

Theorem 2.4.7. *Let D be a UFD. Then $D[x]$ is a UFD.*

Proof. Let $f(x) \in D[x]$, where $f(x)$ is neither 0 nor a unit.

If $f(x)$ is of degree 0, we are done, since D is a UFD. Suppose that
$\deg f(x) > 0$. Let

$$f(x) = g_1(x)g_2(x) \cdots g_r(x)$$

be a factorization of $f(x)$ in $D[x]$ having the greatest number r of
factors of positive degree. Now write each $g_i(x) = a_i h_i(x)$ where a_i
is a content of $g_i(x)$ and $h_i(x)$ is a primitive polynomial. From the
maximality of r, each of the $h_i(x)$ is irreducible. Thus we now have

$$f(x) = a_1 a_2 \cdots a_r h_1(x) h_2(x) \cdots h_r(x)$$

where the $h_i(x)$'s are irreducibles in $D[x]$. If we now factor the
$a_1 a_2 \cdots a_r$ into irreducibles in D, we obtain a factorization of $f(x)$
into a product of irreducibles in $D[x]$.

Now we prove the uniqueness. Let

$$a_1 a_2 \cdots a_r g_1(x) g_2(x) \cdots g_s(x) = b_1 b_2 \cdots b_{r'} h_1(x) h_2(x) \cdots h_{s'}(x)$$
$$(2.2)$$

where the a_i, b_i, $g_j(x)$, $h_j(x)$ are irreducibles in $D[x]$. Then $a_1 a_2 \cdots a_r$
$\sim b_1 b_2 \cdots b_{r'}$ since they are content of the above polynomial, and also

$$g_1(x) g_2(x) \cdots g_s(x) \sim h_1(x) h_2(x) \cdots h_s(x) \text{ in } F[x].$$

Then after renumbering b_i's and using Theorem 2.4.5 and the fact
that $F[x]$ is a UFD, we have

$$r = r', \ a_i \sim b_i \text{ in } D,$$

$$s = s', \ g_j(x) \sim h_j(x) \text{ in } F[x].$$

Note that $g_j(x), h_j(x)$ are irreducibles in $F[x]$. There are $c_j, d_j \in D^*$
such that $g_j(x) = \frac{c_j}{d_j} h_j(x)$. Then $d_j g_j(x) = c_j h_j(x)$ and further
$c_j \sim d_j$ in D, hence $g_j(x) \sim h_j(x)$ in $D[x]$. The uniqueness follows.
□

Corollary 2.4.8. *Let F be a field and $x_1, \cdots, x_n$ indeterminates. Then $F[x_1, \cdots, x_n]$ is a UFD.*

Proof. By the above theorem, $F[x_1]$ is a UFD. Again by the above theorem, so is $(F[x_1])[x_2] = F[x_1, x_2]$. Continuing in this procedure, by induction we see that $F[x_1, \cdots, x_n]$ is a UFD. $\qquad\square$

We have seen that a PID is a UFD. It is easy for us to give an example that shows that not every UFD is a PID.

Example 2.4.3. *Consider the polynomial ring $F[x, y]$ over a field F. We know that $F[x, y]$ is a UFD. Let $I = xF[x, y] + yF[x, y]$. Then $I \lhd F[x, y]$. If $I = aF[x, y]$ for some $a \in F[x, y]$. Since $x, y \in I$, we have $a|x, a|y$. Thus $a \in F$ which is impossible. Hence $F[x, y]$ is not a PID.*

Now we have following generalization of Theorem 1.7.9.

Theorem 2.4.9 (Schönemann-Eisenstein Criterion). *Let D be a UFD with quotient field F, and $f(x) = a_n x^n + \cdots + a_1 x + a_0 \in D[x]$ with $n \geq 1$ and $a_n \neq 0$. If p is prime in D such that*

(i). $p|a_i$ for $0 \leq i < n$,

(ii). $p \nmid a_n$,

(iii). $p^2 \nmid a_0$,

then $f(x)$ is irreducible over F.

Proof. The proof is identical to that of Theorem 1.7.9.

For a contradiction, suppose that $f(x)$ is reducible in $F[x]$, and $f(x) = g(x)h(x)$ for some positive degree polynomials $g(x), h(x) \in F[x]$. By Corollary 2.4.6 we may assume that $g(x), h(x) \in D[x]$ as well. Then denote by $\overline{f}(x), \overline{g}(x), \overline{h}(x)$ the reductions mod p of these polynomials, i.e., consider them as polynomials with coefficients in the ID $D/\langle p \rangle$ (since $\langle p \rangle$ is prime). We have $\overline{g}\overline{h}(x) = \overline{f}(x) = \overline{a}_n x^n$, which means that $\overline{g}(x) = \overline{a}x^k$ and $\overline{h}(x) = \overline{b}x^{n-k}$. This shows that all the other coefficients of g and h are zero mod p, so they're all divisible by p. Since their constant terms are both divisible by p, the constant term of $f(x)$ is divisible by p^2, which contradicts the hypothesis of the theorem. $\qquad\square$

Example 2.4.4. *Show that $f(x, y) = x^3 + y^3 + 1$ is irreducible in $\mathbb{Q}[x, y]$. You need to take $D = \mathbb{Q}[y]$, $p(y) = y + 1$ and use Schönemann-Eisenstein Criterion.*

2.5. Multiplicative norms.

In this section we shall give some examples of Euclidean domains different from the ring $\mathbb{Z}$ of integers and the polynomial ring $F[x]$.

Definition 2.5.1. Let $\mathbb{Z}[i] = \{a + bi : a, b \in \mathbb{Z}\}$ which is a subring of $\mathbb{C}$. Any number in $\mathbb{Z}[i]$ is called a **Gaussian integer**. The **norm** of $a + bi \in \mathbb{Z}[i]$, where $a, b \in \mathbb{Z}$, is define as $N(a + bi) = |a + bi|^2 = a^2 + b^2$.

We can easily extend the function N to $\mathbb{C}$, i.e., define $N(a + bi) = a^2 + b^2$ for any $a + bi \in \mathbb{C}$ where $a, b \in \mathbb{R}$. Note that the Gaussian integers include all the integers. Recall that the norm or absolute value of $a + bi \in \mathbb{C}$, where $a, b \in \mathbb{R}$, was defined as $|a + bi| = \sqrt{a^2 + b^2} = \sqrt{N(a + bi)}$. So here we have different meaning for the word norm.

Lemma 2.5.2. For all $\alpha, \beta \in \mathbb{C}$ we have

   (i). $N(\alpha) \geq 0$,
   (ii). $N(\alpha) = 0$ if and only if $\alpha = 0$,
   (iii). $N(\alpha\beta) = N(\alpha)N(\beta)$.

Proof. These results directly follow from properties of absolute value of complex numbers. □

Lemma 2.5.3. $\mathbb{Z}[i]$ is an integral domain.

Proof. This follows from the fact that $\mathbb{Z}[i] \subset \mathbb{C}$ which is a field. □

Theorem 2.5.4. The norm $N(\alpha)$ for nonzero $\alpha \in \mathbb{Z}[i]$ is a Euclidean norm on $\mathbb{Z}[i]$, i.e., $\mathbb{Z}[i]$ is a Euclidean domain.

Proof. For $\beta = b_1 + b_2 i \neq 0$ we know that $N(b_1 + b_2 i) = b_1^2 + b_2^2$. So $N(\beta) \geq 1$. Then

$$N(\alpha) \leq N(\alpha)N(\beta) = N(\alpha\beta), \forall \alpha, \beta \in \mathbb{Z}[i] \setminus \{0\}.$$

This proves Condition (ii) in Definition 2.3.1 for a Euclidean norm.

Now we prove Condition (i) in Definition 2.3.1 for N. Let $\alpha = a_1 + a_2 i$, $\beta = b_1 + b_2 i \in \mathbb{Z}[i]$, where $\beta \neq 0$. We want to find σ and ρ in $\mathbb{Z}[i]$ such that $\alpha = \beta\sigma + \rho$, where either $\rho = 0$ or $N(\rho) < N(\beta) = b_1^2 + b_2^2$.

Let $\frac{\alpha}{\beta} = r + si$ for $r, s \in \mathbb{Q}$. Take $q_1, q_2 \in \mathbb{Z}$ such that $|r - q_1| \leq 1/2$ and $|s - q_2| \leq 1/2$. Let $\sigma = q_1 + q_2 i$ and $\rho = \alpha - \beta\sigma$. If $\rho = 0$, we are done. Otherwise, we see that

$$N\left(\frac{\alpha}{\beta} - \sigma\right) = N((r + si) - (q_1 + q_2 i))$$

$$= N((r - q_1) + (s - q_2)i) \leq (1/2)^2 + (1/2)^2 = 1/2.$$

Thus we obtain

$$N(\rho) = N(\alpha - \beta\sigma) = N\left(\beta\left(\frac{\alpha}{\beta} - \sigma\right)\right) = N(\beta)N\left(\frac{\alpha}{\beta} - \sigma\right)$$

$$\leq N(\beta)/2 < N(\beta).$$

$\square$

Example 2.5.1. *In $\mathbb{Z}[i]$, find $\mathcal{U}(\mathbb{Z}[i])$ and factor 5 into a product of irreducibles.*

Solution. In $\mathbb{Z}[i]$, since $N(1) = 1$, the units of $\mathbb{Z}[i]$ are exactly the $\alpha = a_1 + a_2i$ with $N(\alpha) = a_1^2 + a_2^2 = 1$. Since $a_1, a_2 \in \mathbb{Z}$, it follows that $a_1 = \pm 1$ with $a_2 = 0$, or $a_1 = 0$ with $a_2 = \pm 1$. Thus $\mathcal{U}(\mathbb{Z}[i]) = \{\pm 1, \pm i\}$.

We know that 5 is an irreducible in $\mathbb{Z}$. But 5 is no longer an irreducible in $\mathbb{Z}[i]$ since $5 = (1 + 2i)(1 - 2i)$, where neither $1 + 2i$ nor $1 - 2i$ is a unit. $\square$

Example 2.5.2. *Use a Euclidean algorithm in $\mathbb{Z}[i]$ to find a $\gcd(8 + 6i, 5 - 15i)$.*

Solution. Since $\frac{5-15i}{8+6i} = -\frac{1}{2} - \frac{3}{2}i$, we have $5 - 15i = -i(8 + 6i) - (1 + 7i)$. Since $\frac{8+6i}{1+7i} = 1 - i$, we have $8 + 6i = (1 + 7i)(1 - i)$. We put them together

$$5 - 15i = -i(8 + 6i) - (1 + 7i),$$
$$8 + 6i = (1 + 7i)(1 - i) + 0.$$

Thus $\gcd(8 + 6i, 5 - 15i) \sim 1 + 7i$. $\square$

Let us study integral domains that have a multiplicative norm.

Definition 2.5.5. *A **multiplicative norm** N on an integral domain D is a function mapping $N : D \to \mathbb{Z}$ with the following conditions hold for all $\alpha, \beta \in D$.*

(i). $N(\alpha) = 0$ if and only if $\alpha = 0$.
(ii). $N(\alpha\beta) = N(\alpha)N(\beta)$.

Note that the Euclidean norm on $\mathbb{C}[x]$ is no longer a multiplicative norm.

Theorem 2.5.6. *Let D be an ID with a multiplicative norm N.*

(i). $|N(u)| = 1$ for every unit $u \in \mathcal{U}(D)$.

(ii). *If* $\mathcal{U}(D) = \{\alpha \in D : |N(\alpha)| = 1\}$ *and* $\beta \in D$ *is such that* $|N(\beta)| = p$ *for a prime* $p \in \mathbb{Z}$, *then* β *is an irreducible of* D.

Proof. (i). From

$$N(1) = N((1)(1)) = N(1)N(1)$$

we see that $N(1) = 1$. If u is a unit in D, then

$$1 = N(1) = N(uu^{-1}) = N(u)N(u^{-1}).$$

Since $N(u)$ is an integer, we deduce t that $|N(u)| = 1$.

(ii). If $\beta = \alpha\gamma$ where $\alpha, \gamma \in D$, we have

$$p = |N(\beta)| = |N(\alpha)N(\gamma)| = |N(\alpha)| \cdot |N(\gamma)|.$$

Then either $|N(\alpha)| = 1$ or $|N(\gamma)| = 1$. By (i) we know that either α or γ is a unit of D. So β is an irreducible of D. $\qquad\square$

Example 2.5.3. *It is easy to see that the function* N *defined by* $N(a + bi) = a^2 + b^2$ *gives a multiplicative norm on* $\mathbb{Z}[i]$. *We know that*

$$\mathcal{U}(\mathbb{Z}[i]) = \{\alpha \in \mathbb{Z}[i] : |N(\alpha)| = 1\} = \{\pm 1, \pm i\}.$$

We see that 13 *is not an irreducible in* $\mathbb{Z}[i]$ *since* $13 = (3+2i)(3-2i)$. *Since* $N(3 + 2i) = N(3 - 2i) = 3^2 + 2^2 = 13$ *and* 13 *is a prime in* $\mathbb{Z}$, *we see from the above theorem that* $3 + 2i$ *and* $3 - 2i$ *are both irreducibles in* $\mathbb{Z}[i]$.

The next example gives another example of an integral domain that is not a UFD.

Example 2.5.4. *We now consider the ID*

$$\mathbb{Z}[\sqrt{-5}] = \{a + b\sqrt{-5}\,i : a, b \in \mathbb{Z}\} \subset \mathbb{C}.$$

Define a multiplicative norm N *on* $\mathbb{Z}[\sqrt{-5}]$ *by*

$$N(a + b\sqrt{-5}) = a^2 + 5b^2, \forall a, b \in \mathbb{Z}.$$

Clearly, $N(a + b\sqrt{-5}) = 0$ *if and only if* $a^2 + 5b^2 = 0$ *if and only if* $a = b = 0$ *if and only if* $a + b\sqrt{-5} = 0$. *It is easy to see that* $N(\alpha\beta) = N(\alpha)N(\beta)$ *for any* $\alpha, \beta \in \mathbb{Z}[\sqrt{-5}]$. *So* N *is a multiplicative norm* N *on* $\mathbb{Z}[\sqrt{-5}]$.

Now $N(a + b\sqrt{-5}) = 1$ *if and only if* $a^2 + 5b^2 = 1$ *if and only if* $a = \pm 1$ *and* $b = 0$ *if and only if* $a + b\sqrt{-5} = \pm 1$. *So* $\mathcal{U}(\mathbb{Z}[\sqrt{-5}]) = \{\pm 1\}$.

In $\mathbb{Z}[\sqrt{-5}]$, we have

$$9 = 3 \cdot 3, \quad 9 = (2 + \sqrt{-5})(2 - \sqrt{-5}).$$

We will show that $3, 2 + \sqrt{-5}$, and $2 - \sqrt{-5}$ are all irreducibles in $\mathbb{Z}[\sqrt{-5}]$. Then $\mathbb{Z}[\sqrt{-5}]$ is not a UFD.
If $3 = \alpha\beta$, then

$$9 = N(3) = N(\alpha)N(\beta).$$

We see that $N(\alpha) = 1, 3$, or 9. If $N(\alpha) = 1$, then α is a unit. If $N(\alpha) = 3$ then $a^2 + 5b^2 = 3$ which is impossible. If $N(\alpha) = 9$, then $N(\beta) = 1$, so β is a unit. Thus 3 is an irreducible in $\mathbb{Z}[\sqrt{-5}]$. A similar argument shows that $2 + \sqrt{-5}$ and $2 - \sqrt{-5}$ are also irreducibles in $\mathbb{Z}[\sqrt{-5}]$. Hence $\mathbb{Z}[\sqrt{-5}]$ is an integral domain but not a UFD.

In conclusion, we know that the numbers

$$\pm 3, 2 + \sqrt{-5}, 2 - \sqrt{-5}$$

are all irreducibles in $\mathbb{Z}[\sqrt{-5}]$, but none of them is prime. □

2.6. Exercises.

(1) Show that the ring $(\mathbb{Z}[x], +, \cdot)$ is not a PID.

(2) If p is an irreducible in a UFD D, show that p is a prime.

(3) Factor the polynomial $4x^2 - 4x + 8$ into a product of irreducibles viewing it as an element of the UFD $\mathbb{Z}[x]$; or $\mathbb{Q}[x]$; or $\mathbb{Z}_{11}[x]$.

(4) Find a gcd of the following polynomials in $\mathbb{Q}[x]$:

$$x^{10} - 3x^9 + 3x^8 - 11x^7 + 11x^6 - 11x^5$$
$$+ 19x^4 - 13x^3 + 8x^2 - 9x + 3,$$

$$x^6 - 3x^5 + 3x^4 - 9x^3 + 5x^2 - 5x + 2.$$

(5) Let $\alpha, \beta \in \mathbb{Z}[i]$. Show that $\gcd(\alpha, \beta) \sim 1$ if and only if $\gcd(N(\alpha), N(\beta)) = 1$, where $N(\alpha) = |\alpha|^2$.

(6) Let D be a UFD, F be the field of quotients of D, $f(x_1, x_2, \cdots, x_n) \in D[x_1, x_2, \cdots, x_n]$ be primitive in the obvious meaning. Show that the polynomial $f(x_1, x_2, \cdots, x_n)$ is irreducible in $D[x_1, x_2, \cdots, x_n]$ if and only if it is irreducible in $F[x_1, x_2, \cdots, x_n]$.

(7) Show that $f(x) = x^4 - 4x^2 + 1$ is irreducible in $\mathbb{Q}[x]$ but it is reducible in $\mathbb{Z}_p[x]$ for any prime p.

(8) Use the Euclidean algorithm and UFD's property in $\mathbb{Z}[i]$ to find $\gcd(15 - 12i, 6 - 5i)$, and $\gcd(16 + 7i, 10 - 5i)$.

(9) Find all prime numbers p such that $p = a^4 + 4b^4$, where $a, b \in \mathbb{Z}$. (Hint: Consider the numbers in $\mathbb{Z}[i]$.)

(10) Prove that $\mathbb{Z}[i]/\langle 2 + i \rangle$ is a field.

(11) Let R be a PID, $a, b \in R$. Prove that $\gcd(a, b) \sim 1$ if and only if there exist $u, v \in R$ such that $au + bv = 1$.

(12) Show that in a PID, any proper ideal is contained in a maximal ideal.

(13) Show that the integer solutions of $x^2 + 2 = y^3$ are $x = \pm 5$, $y = 3$.

Hints: We know that $D = \mathbb{Z}[\sqrt{-2}]$ is a Euclidean domain under the usual complex norm. Let $x, y \in \mathbb{Z}$ such that $x^2 + 2 = y^3$.

(a). Show that $\sqrt{-2}$ is irreducible in D.

(b). Show that $(x + \sqrt{-2})$, $(x - \sqrt{-2})$ are relatively prime.

(c). Deduce that the integer solutions of $x^2 + 2 = y^3$ are $x = \pm 5, y = 3$.

(14) Let D be a PID.

(a). Show that every nonzero prime ideal of D is maximal.

(b). If S is an integral domain and $\phi : D \to S$ is a surjective ring homomorphism, show that either ϕ is an isomorphism or S is a field.

(c). If $D[x]$ is a PID, show that D is a field.

(15) Let $R = \mathbb{Z}[\sqrt{-a}]$, where a is an integer ≥ 3. Show that $2R$ is not a prime ideal in R, but that 2 is an irreducible element of R. Is R a PID? a UFD? Why or why not?

(16) Let D be a PID, and let I, J be nonzero principal ideals of D. Show that $IJ = I \cap J$ if and only if $I + J = D$.

(17) Let F be a field, and $R = \{f(x) \in F[x] : f'(0) = 0\}$.

(a). Show that R is a subring of $F[x]$.

(b). Show that x^2 and x^3 are irreducibles in R.

(c). Show that R is not a UFD.

(d). Explicitly demonstrate an ideal that is not principal.

(18) Show that the following polynomials are irreducible in the integral domain $\mathbb{C}[x, y]$:

$$y^5 + xy^4 - y^4 + x^2y^2 - 2xy^2 + y^2 + x^3 - 1,$$
$$xy^3 + x^2y^2 - x^5y + x^2 + 1.$$

(19) For any integer $n > 2$, show that the following polynomials

$$f(x) = x^n + 2x^{n-1} + 2x^{n-2} + \cdots + 2x + 1 + i,$$

$$g(x) = x^n + 5x^{n-1} + 5x^{n-2} + \cdots + 5x + 2 + i$$

are irreducible over $\mathbb{Z}[i]$.

(20) Show that $x^6 + x^3 + 1$ is irreducible over $\mathbb{Q}$.

(21) Show that the rings $\mathbb{Z}[\sqrt{2}]$ and $\mathbb{Z}[\sqrt{3}]$ are Euclidean domains.

(22) Show that 6 does not factor uniquely in the ID $\mathbb{Z}[\sqrt{-5}]$.

(23) Let F be a field. Find all $f(x) \in F[x]$ such that $f(x^2) = f(x)^2$.

(24) Let $n \in \mathbb{N}$ and $a_1, a_2, \cdots, a_n \in \mathbb{Z}$ be pairwise distinct. Show that

$$f(x) = (x - a_1)(x - a_2) \cdots (x - a_n) - 1$$

is irreducible over $\mathbb{Q}$. (This was proved by Issai Schur (1875–1941) in 1908. Hints: If $f(x) = g(x)h(x)$, then $g(a_i)h(a_i) = -1$, and furthermore $g(a_i) = -h(a_i)$.)

(25) Let $n \in \mathbb{N}$ and $a_1, a_2, \cdots, a_n \in \mathbb{Z}$ be pairwise distinct. Show that

$$f(x) = [(x - a_1)(x - a_2) \cdots (x - a_n)]^2 + 1$$

is irreducible over $\mathbb{Q}$.

(26) Let $f(x) \in \mathbb{Z}[x]$ such that it takes value 1 at four distinct integers. Show that $f(x)$ does not take value -1 at any integer.

(27) Let $g(x) = ax^2 + bx + 1 \in \mathbb{Z}[x]$ be irreducible over $\mathbb{Q}$ of degree 2. Let $n \in \mathbb{N}$ and $a_1, a_2, \cdots, a_n \in \mathbb{Z}$ be pairwise distinct and let

$$f(x) = (x - a_1)(x - a_2) \cdots (x - a_n).$$

Show that $g(f(x))$ is irreducible over $\mathbb{Q}$ if $n \geq 7$.

(28) Show that, the result in Perron's Irreducibility Criterion 1.8.1 still holds if we replace $\mathbb{Z}$ and $\mathbb{Q}$ with $\mathbb{Z}[i]$ and $\mathbb{Q}[i]$, respectively.

(29) Let p be a prime in $\mathbb{Z}[i]$,

$$f(x) = a_n x^n + a_{n-1} x^{n-1} + \cdots + a_1 x + p \in \mathbb{Z}[i][x],$$

with $n \geq 1$ and $a_n \neq 0$. If $|p| > |a_1| + \cdots + |a_n|$, show that $f(x)$ is irreducible over $\mathbb{Q}[i]$.

3. Modules and Noetherian rings

In this chapter we will introduce Noetherian rings and modules over a ring, which are powerful tools to study rings. We will establish very basic properties for Noetherian rings and modules. In this chapter we always assume that R is a ring. Unlike many other books, we do not assume that R is unital at the beginning.

3.1. Modules, submodules and isomorphism theorems.

Definition 3.1.1. *A set M is called a* **left R-module** *or a* **module over** *R if M is an additive abelian group with a map $R \times M \to M$ defined by $(r, u) \to ru$ such that for $u, v \in M$ and $r_1, r_2 \in R$ we have*

(1). $r_1(u + v) = r_1 u + r_1 v$,

(2). $(r_1 + r_2)u = r_1 u + r_2 u$,

(3). $(r_1 r_2)u = r_1(r_2 u)$.

A **right R-module** *can be defined analogously. Here the product of $u \in M$ and $r \in R$ is denoted by ur.*

Example 3.1.1. *(1). R and $\{0\}$ are naturally left (and also right) R-modules in the similar manner. These R-modules are called* **regular left and regular right R-modules**.

(2). Any abelian group $(A, +)$ can be considered a left $\mathbb{Z}$-module as follows. For $g \in A$ and $k \in \mathbb{Z}$ we defined

$$kg = \underbrace{g + \cdots + g}_{k \text{ times}} \text{ if } k > 0, 0_{\mathbb{Z}} g = 0_A,$$

and $kg = -[(-k)g]$ if $k < 0$.

(3). The set $M_n(R)$ of all $n \times n$ matrices over a ring R becomes a left R-module if we define

$$rX = \begin{pmatrix} r & 0 & 0 & \cdots & 0 \\ 0 & r & 0 & \cdots & 0 \\ 0 & 0 & r & \cdots & 0 \\ \vdots & \vdots & \vdots & & \ddots \\ 0 & 0 & 0 & & r \end{pmatrix} X, \forall r \in R \text{ and } X \in M_n(R).$$

Clearly, we can also make $M_n(R)$ a right R-module in the same manner.

By M_R we denote a right R-module M, while $_RM$ will denote M as a left R-module. For convenience, we generally work with left R-modules while dealing with non-commutative rings. We simply say that M is a module if other details are clear from the context.

When R is a commutative ring, left R-module and right R-modules are the same, i.e., if M is a R-module we can use $ru = ur$ for any $r \in R$ and $u \in M$.

Proposition 3.1.2. *Let M be a left R-module. Then:*
  (1). $r0_M = 0_M$ for all $r \in R$,
  (2). $0_R u = 0_M$ for all $u \in M$,
  (3). $r(-u) = (-r)u = -ru$ for all $u \in M$ and $r \in R$.

Proof. This is easy to prove.   □

Definition 3.1.3. *Let M be a left R-module.*
  *(a). A subset K of M is called a R-**submodule** of M if K is also a left R-module under the same action defined on M, denoted by $K \le M$.*
  *(b). A nonzero R-module M is called to be **simple** if $\{0\}$ and M are the only submodules of M.*
  *(c). Let $S \subseteq R$. The submodule of M **generated** by S is the smallest submodule of M that contains S, denoted by $\langle S \rangle$.*

Proposition 3.1.4. *Let K be a non-empty subset of $_RM$. Then $K \le M$ if and only if $u - v \in K$ and $rx \in K$ for all $u, v \in K$ and $r \in R$.*

Proof. This is easy to prove. We omit the details.   □

Definition 3.1.5. *Submodules of R_R are called **right ideals** of R and submodules of $_RR$ are called **left ideals** of R.*

Let K be a submodule of a left R-module M. Consider the factor group M/K. Elements of M/K are cosets of the form $u + K$ with $m \in M$. We can make M/K a left R-module by defining

$$r(u + K) = ru + K \ \forall \ u \in M \text{ and } r \in R.$$

Check that this action is well defined and the module axioms are satisfied to make M/K a left R-module. We define the following sets

$$[\![M, K]\!] = \{L \le M \mid L \supseteq K\}, \quad [\![M/K]\!] = \{K' \le M/K\}.$$

Similar to rings and subrings we have the following correspondence theorem for submodules.

Proposition 3.1.6. *Let K be a submodule of $_RM$.*

(1). Every submodule of M/K has the form A/K where A is a submodule of M and $A \supseteq K$.

(2). The map $[\![M, K]\!] \to [\![M/K]\!]$, $L \mapsto L/K$ is a one to one and onto map.

Definition 3.1.7. *Let M and M' be left R-modules. A map $\varphi : M \to M'$ is called an R-**module homomorphism** if:*

$$\varphi(u + v) = \varphi(u) + \varphi(v), \ \forall u, v \in M,$$
$$\varphi(ru) = r\varphi(u), \ \forall u \in M, \ r \in R.$$

If K is a submodule of $_RM$ then the map

$$\sigma : M \to M/K, \ \sigma(m) = m + K \ \forall m \in M$$

is a homomorphism of M onto M/K. It is called the **canonical homomorphism**.

Proposition 3.1.8. *Let $\varphi :_R M \to_R M'$ be an R-module homomorphism. Then:*

(1). $\varphi(0_M) = 0_{M'}$;

(2). $\ker(\varphi) = \{u \in M : \varphi(u) = 0_{M'}\} \leq M$;

(3). $\varphi(M) = \{\varphi(u) : u \in M\} \leq M'$.

Proof. This is easy to prove. We omit the details.   □

The above $\ker(\varphi)$ is called the **kernel** of φ, and $\varphi(M)$ is called the **image** of φ.

Definition 3.1.9. *Let $\varphi : M \to M'$ be an R-homomorphism. Then φ is called an R-**isomorphism** if it is in addition a one to one and onto map. In this case we write $M \cong M'$.*

There are similar isomorphism theorems to those for rings.

Theorem 3.1.10 (First isomorphism theorem). *Let M and M' be left R-modules and $\varphi : M \to M'$ and R-homomorphism. Then $\varphi(M) \cong M/\ker(\varphi)$.*

Theorem 3.1.11 (Second isomorphism theorem). *Let L, K be submodules of $_RM$. Then $(L + K)/K \cong L/(L \cap K)$.*

Theorem 3.1.12 (Third isomorphism theorem). *If K, L are submodules of $_R M$ and $K \subseteq L$ then $L/K \leq M/K$ and $(M/K)/(L/K) \cong M/L$.*

The proofs of these theorems are similar to those for rings. Let $M_1, \cdots, M_n$ be left R-modules. The set of n-tuples $\{(u_1, \cdots, u_n) : u_i \in M_i\}$ becomes a left R-modules if we define

$$(u_1, \cdots, u_n) + (u_1', \cdots, u_n') = (u_1 + u_1', \cdots, u_n + u_n')$$

and $r(u_1, \cdots, u_n) = (r u_1, \cdots, r u_n)$. This is the **external direct sum** of the M_i and is denoted

$$\oplus_{i=1}^n M_i \text{ or } M_1 \oplus \cdots \oplus M_n.$$

For simplicity we denote $M^n = M \oplus \cdots \oplus M$, the direct sum of n copies of an R-module M. For convenience, sometimes we use column vectors to denote elements in M^n, say $(u_1, \cdots, u_n)^{\mathrm{t}} \in M^n$.

Let $\{M_\lambda\}_{\lambda \in \Lambda}$ be a collection of submodules of a left R-modules M. We define their **sum**

$$\sum_{\lambda \in \Lambda} M_\lambda = \{u_{\lambda_1} + \cdots + u_{\lambda_k} : u_{\lambda_i} \in M_{\Lambda_i} \text{ for all possible subsets}$$

$$\{\lambda_1, \cdots, \lambda_k\} \text{ of } \Lambda\}.$$

Thus $\sum_{\lambda \in \Lambda} M_\lambda$ is the set of all finite sums of elements of the M_λ's. It is easy to check that this is a submodule of M.

$\sum_{\lambda \in \Lambda} M_\lambda$ is said to be **direct** if each element in $\sum_{\lambda \in \Lambda} M_\lambda$ has a unique expression as $u_{\lambda_1} + \cdots + u_{\lambda_k}$ for some $u_{\lambda_i} \in M_{\lambda_i}$. As before we can show that

$$\sum_{\lambda \in \Lambda} M_\lambda \text{ is direct} \iff M_\mu \cap \left\{ \sum_{\lambda \in \Lambda, \lambda \neq \mu} M_\lambda \right\} = \{0\} \; \forall \; \mu \in \Lambda.$$

If $\sum_{\lambda \in \Lambda} M_\Lambda$ is direct and Λ is a finite set, we denote it by $\oplus_{i=1}^n M_i$ or $M_1 \oplus \cdots \oplus M_n$. As explained for rings before, there is no real difference between (finite) external and internal direct sums of modules.

Definition 3.1.13. *Let R be a unital ring. A module $_R M$ is said to be **unital** if $1u = u$ for all $u \in M$.*

We shall assume that all modules considered are unital whenever R is a unital ring. A vector space V over a field F is exactly a unital F-module.

Definition 3.1.14 (Cyclic submodule). *A submodule of an R-module M generated by a single nonzero element is called a* **cyclic submodule** *of M. An R-module M is call to be* **cyclic** *if it can be generated by a single element.*

Definition 3.1.15 (Finitely generated module). *An R-module M is* **finitely generated** *if there are $u_1, \cdots, u_n \in M$ such that $M = \langle u_1, \cdots, u_n \rangle$.*

Let R be a ring with an ideal I and M a left R-module. In general, M need not be a left R/I-module. However, we can give M a left R/I-module structure if $IM = 0$. In this case we define

$$(r + I)u = ru \text{ for all } u \in R \text{ and } r \in R.$$

It can be checked that this is a well-defined left R/I-module action. Further, under this action the R-submodules and R/I-submodules of M coincide.

Example 3.1.2. *Let I, J be ideals of the unital ring R. Show that $R/I \cong R/J$ as left R-modules if and only if $I = J$.*

Proof. If $I = J$ we clearly have $R/I \cong R/J$ as left R-modules.

Now suppose that $R/I \cong R/J$ as left R-modules. Since $I(R/I) = 0$ we have $I(R/J) = 0$, i.e., $I \subset J$. Similarly $J \subset I$. Thus $I = J$. □

3.2. Free modules.

Let R be a unital ring and M a left R-module.

Definition 3.2.1 (Linear independence). *Let $u_1, \cdots, u_n \in M$. Then $\{u_1, \cdots, u_n\}$ is* **linearly independent** *if*

$$\sum_{i=1}^{n} r_i u_i = 0 \text{ for } r_i \in R$$

implies $r_1 = r_2 = \cdots = r_n = 0$.

Definition 3.2.2 (Freely generate). *A subset $S \subseteq M$ generates M* **freely** *if*

(1) S generates M, and
(2) any set map $\varphi : S \to N$ to an R-module N can be extended to an R-module homomorphism $\tilde{\varphi} : M \to N$.

One can show that the R-module homomorphism $\tilde{\varphi} : M \to N$ is uniquely determined by the map φ. See Exercise (15).

Thus, what this definition tells us is that giving an R-module homomorphism from M to N is exactly the same thing as giving a map from S to N.

Definition 3.2.3 (Free module and basis). *An R-module M is **free** if it is freely generated by some subset $S \subseteq M$, and S is called a* **basis** *of M.*

Similar to what we do in linear algebra, we have

Proposition 3.2.4. *Let M be a module over a unital ring R. For a subset $S = \{u_1, \cdots, u_n\} \subseteq M$, the following are equivalent:*

(a). S is a basis of M;
(b). S generates M and S is linearly independent;
(c). Every element of $u \in M$ is uniquely expressible as

$$u = r_1 u_1 + r_2 u_2 + \cdots + r_n u_n$$

for some $r_i \in R$.

Proof. The proof for the equivalence of (b) and (c) is the same as in Linear Algebra. So we only show that (a) and (b) are equivalent.

(a)$\Rightarrow$(b). If S is not independent, then we can write

$$r_1 u_1 + \cdots + r_n u_n = 0,$$

with $r_i \in M$ and, say, r_1 non-zero. We define the set function $\varphi : S \to R$ by sending $u_1 \mapsto 1_R$ and $u_i \mapsto 0$ for all $i \neq 1$. As S generates M freely, this extends to an R-module homomorphism $\tilde{\varphi} : M \to R$.

By definition of a homomorphism, we can compute

$$
\begin{aligned}
0 &= \tilde{\varphi}(0) \\
&= \tilde{\varphi}(r_1 u_1 + r_2 u_2 + \cdots + r_n u_n) \\
&= r_1 \tilde{\varphi}(u_1) + r_2 \tilde{\varphi}(u_2) + \cdots + r_n \tilde{\varphi}(u_n) \\
&= r_1.
\end{aligned}
$$

This is a contradiction. So (b) follows.

(b)$\Rightarrow$(a). Suppose every element can be uniquely written as $r_1 u_1 + \cdots + r_n u_n$. Given an R-module N any set function $\varphi : S \to N$, we define $\tilde{\varphi} : M \to N$ by

$$\tilde{\varphi}(r_1 u_1 + \cdots + r_n u_n) = r_1 \varphi(u_1) + \cdots + r_n \varphi(u_n).$$

This is well-defined by uniqueness, and is clearly a homomorphism. So it follows that S is a basis of M. $\square$

Proposition 3.2.5. *Let M be a module over a unital ring R. Let $S = \{u_1, \cdots, u_n\}$ be a basis for the R-module M. Then $M \cong R^n$.*

Proof. We can easily prove that the map

$$\varphi : R^n \to M, (r_1, r_2, \cdots, r_n) \mapsto r_1 u_1 + \cdots + r_n u_n$$

is an R-module isomorphism. $\square$

Proposition 3.2.6. *If $I \trianglelefteq R$ is an ideal and M is an R-module, then*
(a). $IM = \{\sum_{i=1}^n a_i u_i : a_i \in I, u_i \in M\} \leq M$;
(b). M/IM is an R/I module via $(r + I) \cdot (u + IM) = r \cdot u + IM$ for all $r \in R, u \in M$.

Proof. (a). It is easy to see that $IM \leq M$.
(b). If $b \in I$, then its action on M/IM is

$$b(u + IM) = bu + IM = IM, \forall u \in M,$$

i.e., everything in I kills everything in M/IM. We can consider M/IM as an R/I module by

$$(r + I) \cdot (u + IM) = r \cdot u + IM, \forall r \in R, u \in M.$$

$\square$

We next need to use the following general fact:

Proposition 3.2.7. *Every unital ring has a maximal ideal.*

Proof. We observe that an ideal $I \triangleleft R$ is proper if and only if $1 \notin I$. So every increasing union of proper ideals is proper. By Kuratowski-Zorn Lemma, there is a maximal ideal of R. $\square$

Lemma 3.2.8. *Let $M_1, M_2, \cdots, M_n$ be R-modules with $H_1 \leq M_1$, $H_2 \leq M_2, \cdots, H_n \leq M_n$. Then*

$$\frac{M_1 \oplus M_2 \oplus \cdots \oplus M_n}{H_1 \oplus H_2 \oplus \cdots \oplus H_n} \cong \frac{M_1}{H_1} \oplus \frac{M_2}{H_2} \oplus \cdots \oplus \frac{M_n}{H_n}.$$

Proof. Consider the map

$$\phi : M_1 \oplus M_2 \oplus \cdots \oplus M_n \to \frac{M_1}{H_1} \oplus \frac{M_2}{H_2} \oplus \cdots \oplus \frac{M_n}{H_n},$$

$$(g_1, g_2, \cdots, g_n) \mapsto (g_1 + H_1, g_2 + H_2, \cdots, g_n + H_n).$$

It is easy to see that ϕ is a module homomorphism with $\ker(\phi) = H_1 \oplus H_2 \oplus \cdots \oplus H_n$. Using the first isomorphism theorem we obtain the result in the lemma. $\square$

Theorem 3.2.9 (Invariance of rank). *Let R be a unital commutative ring. If $R^n \cong R^m$ as R-modules, then $n = m$.*

Proof. Let I be a maximal ideal of R. Suppose we have $R^n \cong R^m$. Then we must have
$$R^n/IR^n \cong R^m/IR^m,$$
as R/I modules. Using the fact that
$$IR^n = (IR)^n, IR^m = (IR)^m, R^n/(IR)^n \cong (R/I)^n,$$
and Lemma 3.2.8 we see that
$$(R/I)^n \cong (R/I)^m,$$
are vector spaces over the field R/I. By Linear Algebra we must have $m = n$. □

Definition 3.2.10. *Let R be a unital commutative ring. If R-module $M \cong R^k$, we say that M is of* **rank** k.

Example 3.2.1. *Let R be a unital commutative ring. If every submodule of a free R-module is free, show that R is a principal ideal domain.*

Proof. We first show that R has no 0-divisors. Otherwise assume that $ab = 0$ for $a, b \in R \setminus \{0\}$. Then the principal ideal Rb is a free module. This is impossible since $a(Rb) = 0$, i.e., Rb cannot have a basis.

Let $I \trianglelefteq R$ be nonzero. Then I is an R-module. From the assumptions we know that I is a free R-module with a basis $\{u_i : i \in J\}$. If $|J| > 1$, say $1, 2 \in J$. We know that $Ru_1 \cap Ru_2 = 0$. Since $u_1 u_2 \in Ru_1 \cap Ru_2$, we deduce that $u_1 u_2 = 0$ which is impossible. We obtain that $|J| = 1$, i.e., $I = Ru$ for some $u \in R$. Thus R is a principal ideal domain. □

3.3. Finitely generated modules over Euclidean domains.

In this section we will study finitely generated modules over a Euclidean domain D.

Theorem 3.3.1. *Let $X = \{x_1, \cdots, x_r\}$ be a basis for a free D-module M and $t \in D$. Then for $i \neq j$, the set*
$$Y = \{x_1, \cdots, x_{j-1}, x_j + tx_i, x_{j+1}, \cdots, x_r\}$$
is also a basis for M.

Proof. Since $x_j = (-t)x_i + (1)(x_j + tx_i)$, we see that $x_j \in \langle Y \rangle$, i.e., Y generates M. Let

$$a_1 x_1 + \cdots + a_{j-1} x_{j-1} + a_j (x_j + tx_i) + a_{j+1} x_{j+1} + \cdots + a_r x_r = 0,$$

where $a_i \in D$. Then

$$a_1 x_1 + \cdots + (a_i + a_j t) x_i + \cdots + a_j x_j + \cdots + a_r x_r = 0.$$

Since X is a basis, we deduce that

$$a_1 = \cdots = a_i + a_j t = \cdots = a_j = \cdots = a_r = 0.$$

We see that

$$a_1 = \cdots = a_i = \cdots = a_j = \cdots = a_r = 0.$$

So Y is a basis. $\qquad\square$

Theorem 3.3.2. *Let D be a Euclidean domain with Euclidean norm $\phi : R \setminus \{0\} \to \mathbb{Z}^+$, M be a nonzero free D-module of finite rank n, and let K be a nonzero submodule of M.*

(1). K is a free D-module of rank $s \le n$;

(2). There exists a basis $\{x_1, x_2, \cdots, x_n\}$ for M and $d_1, d_2, \cdots, d_s \in D$ with $d_i | d_{i+1}$, such that $\{d_1 x_1, d_2 x_2, \cdots, d_s x_s\}$ is a basis for K.

Proof. We only prove (2) since (1) follows from (2).

For any basis $Y = \{y_1, \cdots, y_n\}$ of M, all nonzero elements in K can be expressed in the form

$$a_1 y_1 + \cdots + a_n y_n, a_i \in D$$

where some a_i is nonzero.

Step 1: Constructing x_1.

Among all bases Y for M, select one Y_1 so that $\phi(a_i)$ is minimal as all nonzero elements of K are written in terms of the basis elements in Y_1. By renumbering the elements of Y_1 if necessary, we can assume there is $w_1 \in K$ such that

$$w_1 = d_1 y_1 + a_2 y_2 + \cdots + a_n y_n$$

where $d_1 \ne 0$ and $\phi(d_1)$ is the minimal as just described. Write $a_j = d_1 q_j + r_j$ where $r_j = 0$ or $\phi(r_j) < \phi(d_1)$ for $j = 2, \cdots, n$. Then

$$w_1 = d_1(y_1 + q_2 y_2 + \cdots + q_n y_n) + r_2 y_2 + \cdots + r_n y_n.$$

Take $x_1 = y_1 + q_2 y_2 + \cdots + q_n y_n$. By Theorem 3.3.1 then $X_1 = \{x_1, y_2, \cdots, y_n\}$ is also a basis for M. From our choice of y_1 for minimal coefficient d_1, we see that $r_2 = \cdots = r_n = 0$. Thus $d_1 x_1 \in K$.

Claim 1. If $w = d_1 x_1 + a_2 y_2 + \cdots + a_n y_n \in K$ then $d_1 | a_j$ for any $j = 2, \cdots, n$.

Write $a_j = d_1 q_j + r_j$ where $r_j = 0$ or $\phi(r_j) < \phi(d_1)$ for $j = 2, \cdots, n$. Then

$$w = d_1(x_1 + q_2 y_2 + \cdots + q_n y_n) + r_2 y_2 + \cdots + r_n y_n.$$

Now let $x_1' = x_1 + q_2 y_2 + \cdots + q_n y_n$. By Theorem 3.3.1 then $\{x_1', y_2, \cdots, y_n\}$ is also a basis for M. From our choice of y_1 for minimal coefficient d_1, we see that $r_2 = \cdots = r_n = 0$. Claim 1 follows.

Step 2: Constructing x_2.

We will use the basis $X_1 = \{x_1, y_2, \cdots, y_n\}$. If $a_1 x_1 + a_2 y_2 + \cdots + a_n y_n \in K$ can imply that $a_2 = a_3 = \cdots = a_n = 0$, then $K = \langle d_1 x \rangle$. We are done in this case.

Consider elements of the form

$$a_2 y_2 + \cdots + a_n y_n \in K.$$

Note that $d_1 | a_j$. There is an element in K with minimal $\phi(a_i)$ and $a_i \neq 0$ for some $i = 2, 3, \cdots, n$. By renumbering the elements of X_1 we can assume that there is $w_2 \in K$ such that

$$w_2 = d_2 y_2 + \cdots + a_n y_n$$

where $d_2 \neq 0$ and $\phi(d_2)$ is minimal as just described. Exactly as in Step 1 and Claim 1, we can modify our basis from $X_1 = \{x_1, y_2, \cdots, y_n\}$ to a basis $X_2 = \{x_1, x_2, y_3, \cdots, y_n\}$ for M where $d_1 x_1, d_2 x_2 \in K$, and $w = a_1 x_1 + a_2 y_2 + \cdots + a_n y_n \in K$ implies $d_1 | a_1$ and $d_2 | a_i$ for any $i = 2, \cdots, n$.

Step 3: Finishing.

We have the basis $X_2 = \{x_1, x_2, y_3, \cdots, y_n\}$ for M and examine elements of K of the form $a_3 y_3 + \cdots + a_n y_n$. The pattern is clear. The process continues until we obtain a basis $\{x_1, x_2, \cdots, x_s, y_{s+1}, \cdots, y_n\}$ where the only element of K of the form $a_{s+1} y_{s+1} + \cdots + a_n y_n$ is zero, that is, all a_i are zero. We then let $x_{s+1} = y_{s+1}, \cdots, x_n = y_n$ and obtain a basis for M of the form described in the statement of the theorem. $\square$

Similar to the case in Linear Algebra, we can define the three types of elementary row operations on a matrix $A \in M_{m \times n}(R)$ where R is a commutative unital ring:

1. Multiply a row by a constant $c \in U(R)$. $(cR_i \to R_i)$
2. Interchange two rows. $(R_i \leftrightarrow R_j$ for $i \neq j)$
3. Add a constant c times one row to another. $(cR_i + R_j \to R_j$ for $i \neq j)$

If B is the matrix that results from A by performing one of the above operations, then the matrix A can be recovered from B by performing the corresponding operation in the following list:

1. Multiply the same row by $1/c$. $(c^{-1}R_i \to R_i)$
2. Interchange the same two rows. $(R_i \leftrightarrow R_j$ for $i \neq j)$
3. If B resulted by adding c times row r_i of A to row r_j, then add $-c$ times r_i to r_j. $(-cR_i + R_j \to R_j$ for $i \neq j)$

It follows that if B is obtained from A by performing a sequence of elementary row operations, then there is a second sequence of elementary row operations, which when applied to B recovers A.

Definition 3.3.3. *A matrix $E \in M_n(R)$ is called an* **elementary matrix** *over a commutative unital ring R if it can be obtained from an identity matrix by performing a single elementary row operation.*

Definition 3.3.4 (Invertible matrices). *Let R be a commutative unital ring. A matrix $A \in M_n(R)$ is* **invertible** *if there is $A^{-1} \in M_n(R)$ such that*

$$AA^{-1} = A^{-1}A = I_n.$$

We denote the set of all invertible $n \times n$ matrices in $M_n(R)$ by $\mathrm{GL}_n(R)$ which is called the **general linear group** *of rank n over R.*

Using adjoint matrix of A, one can easily show that $A \in M_n(R)$ is invertible if and only if $\det(A) \in U(R)$. Unlike in Liner Algebra, now $\det(A) \neq 0$ cannot generally deduce the invertibility of A.

Definition 3.3.5 (Equivalent matrices). *Two $m \times n$ matrices A, B over a commutative unital ring R are* **equivalent** *if $B = PAQ$ for some invertible matrices $P \in \mathrm{GL}_m(R)$ and $Q \in \mathrm{GL}_n(R)$.*

We will prove the following important result.

Theorem 3.3.6 (Smith normal form). *An $m \times n$ matrix A over a Euclidean domain D is equivalent to a matrix of the form*

$$\begin{bmatrix} \begin{bmatrix} d_1 & & & \\ & d_2 & & \\ & & \ddots & \\ & & & d_r \end{bmatrix} & O_1 \\ O_2 & O_3 \end{bmatrix}$$

with all the $d_i \in D$ non-zero and $d_1 \mid d_2, d_2 \mid d_3, \cdots d_{r-1} \mid d_r$, where O_1, O_2, O_3 are zero matrices.

Before proving this theorem we will generalize some results from Linear algebra. Assume that V, W are free D-modules.

Definition 3.3.7. *A map $L : V \to W$ is called a* **linear map** *from V to W if, for all $u, v \in V$ and $c \in D$, we have*

(a). $L(u + v) = L(u) + L(v)$,
(b). $L(cu) = cL(u)$.

We often simply call L **linear.** *Linear maps from V into itself are also called a* **linear operators** *on V. We denote the set of all linear maps from V to W by $\mathcal{L}(V, W)$, or by $\mathrm{Hom}_D(V, W)$. In the case that $V = W$, we simply write $\mathcal{L}(V, V)$ as $\mathcal{L}(V)$ which is also denoted by $\mathrm{End}_D(V)$.*

Definition 3.3.8. *An* **ordered basis** *for a free D-module V is a basis for V with a specific order. If $\beta = \{u_1, u_2, \cdots, u_n\}$ is an ordered basis for V, for convenience we simply write it as $\beta = (u_1, u_2, \cdots, u_n)$ and consider it as a $1 \times n$ matrix. So an ordered basis is a $1 \times n$ matrix of vectors.*

We call $(e_1, e_2, \cdots, e_n)$ the **standard ordered basis** for the module D^n where $e_i = (\delta_{1,i}, \delta_{2,i}, \cdots, \delta_{n,i})^{\mathrm{t}}$ and we consider

$$D^n = \left\{ (a_1, a_2, \cdots, a_n)^{\mathrm{t}} : a_i \in D \right\}.$$

Theorem 3.3.9. *Let $\beta = (v_1, v_2, \cdots, v_n)$ be an ordered basis for a free D-module V. For $A, B \in M_{n \times m}(D)$, if $\beta A = \beta B$, then $A = B$.*

Proof. Let $A = (a_{ij})$ and $B = (b_{ij})$. From $\beta A = \beta B$ we see that

$$a_{1i} v_1 + a_{2i} v_2 + \cdots + a_{ni} v_n = b_{1i} v_1 + b_{2i} v_2 + \cdots + b_{ni} v_n.$$

Since $v_1, v_2, \cdots, v_n$ are linearly independent, we deduce that $a_{ij} = b_{ij}$ for all i, j, that is $A = B$.   □

Suppose that $\beta = (v_1, v_2, \cdots, v_n)$ and $\gamma = (w_1, w_2, \cdots, w_m)$ are ordered bases for V and W, respectively. Let $L \in \mathcal{L}(V, W)$. Then there exist unique scalars $a_{ij} \in D$ for each $1 \leq i, j \leq m$, such that

$$
\begin{aligned}
L(v_1) &= a_{11}w_1 + a_{21}w_2 + \cdots + a_{m1}w_m \\
L(v_2) &= a_{12}w_1 + a_{22}w_2 + \cdots + a_{m2}w_m \\
&\cdots\cdots\cdots\cdots \\
L(v_n) &= a_{1n}w_1 + a_{2n}w_2 + \cdots + a_{mn}w_m.
\end{aligned} \tag{3.1}
$$

We define the matrix

$$
A = \begin{bmatrix}
a_{1,1} & a_{1,2} & \cdots & a_{1,n} \\
a_{2,1} & a_{2,2} & \cdots & a_{2,n} \\
\vdots & \vdots & \ddots & \vdots \\
a_{m,1} & a_{m,2} & \cdots & a_{m,n}
\end{bmatrix}.
$$

Definition 3.3.10. *We call the above $m \times n$ matrix $A = (a_{ij})$ the* **matrix** *of L in the ordered bases β and γ and denote $A = [L]_\beta^\gamma$.*

We can write the formula as matrix product

$$
\begin{aligned}
L(\beta) &= (L(v_1), L(v_2), \cdots, L(v_n)) \\
&= (w_1, w_2, \cdots, w_m)[L]_\beta^\gamma = \gamma[L]_\beta^\gamma.
\end{aligned} \tag{3.2}
$$

Note that many properties for linear maps in Linear Algebra trivially hold here for D-linear maps also. For example L is invertible if and only if the matrix $[L]_\beta^\gamma$ is invertible.

Proof of Theorem 3.3.6. Let α, β be the standard ordered basis for D^n, D^m respectively. Consider the D-linear map

$$
L_A : D^n \to D^m, \quad \begin{bmatrix} a_1 \\ a_2 \\ \vdots \\ a_n \end{bmatrix} \mapsto A \begin{bmatrix} a_1 \\ a_2 \\ \vdots \\ a_n \end{bmatrix}.
$$

Then $[L_A]_\alpha^\beta = A$, $L_A(\alpha) = \beta A$ and $L_A(D^n) \leq D^m$. Using Theorem 3.3.2 we have an ordered basis $\beta' = (x_1, x_2, \ldots, x_m)$ for D^m and $d_1, d_2, \ldots, d_s \in D$ with $d_i | d_{i+1}$, such that $(d_1 x_1, d_2 x_2, \ldots, d_s x_s)$ is a basis for $L_A(D^n)$.

There are $y_1, y_2, \ldots, y_r \in D^n$ such that $L_A(y_i) = d_i x_i$. Let $(y_{r+1}, \ldots, y_s)$ be an ordered basis for $\ker(L_A)$. We can easily show that

$\alpha' = (y_1, y_2, \ldots, y_s)$ is a basis for D^n. So $s = n$ and

$$[L_A]_{\alpha'}^{\beta'} = \left[\begin{array}{c|c} \begin{bmatrix} d_1 & & & \\ & d_2 & & \\ & & \ddots & \\ & & & d_r \end{bmatrix} & O_1 \\ \hline O_2 & O_3 \end{array}\right].$$

Using Linear Algebra formulas (say, [Z, Theorem 4.23]) we have

$$[L_A]_{\alpha'}^{\beta'} = [I_{D^n}]_{\beta}^{\beta'} [L_A]_{\alpha}^{\beta} [I_{D^m}]_{\alpha'}^{\alpha} = PAQ$$

where $P = [I_{D^n}]_{\beta}^{\beta'} \in \mathrm{GL}_m(D), Q = [I_{D^m}]_{\alpha'}^{\alpha} \in \mathrm{GL}_m(D)$. $\qquad\square$

We can actually find the Smith normal form of an $m \times n$ matrix A over a Euclidean domain D by simply applying elementary row and column operations to A, see Example 3.3.1.

Definition 3.3.11 (Invariant factors). *The elements $d_k \in D$ obtained in the Smith normal form of the $m \times n$ matrix A are called the* **invariant factors** *of A.*

Theorem 3.3.12 (Classification of finitely-generated modules over a Euclidean domain). *Let D be a Euclidean domain, and M be a finitely generated D-module. Then*

$$M \cong D/\langle d_1 \rangle \oplus D/\langle d_2 \rangle \oplus \cdots \oplus D/\langle d_r \rangle \oplus D^n$$

for some nonzero nonunit $d_i \in D$ with $d_1 \mid d_2, d_2 \mid d_3, \cdots d_{r-1} \mid d_r$.

Proof. Since M is finitely-generated, there is a surjective module homomorphism $\phi : D^m \to M$. So by the first isomorphism, we have

$$M \cong D^m / \ker \phi.$$

Since $\ker \phi$ is a submodule of D^m, by the previous theorem, there is a basis $v_1, \cdots, v_m$ of D^m such that $\ker \phi$ is generated by $d_1 v_1, \cdots, d_r v_r$ for some nonzero $d_i \in D$ with $d_1 \mid d_2, d_2 \mid d_3, \cdots d_{r-1} \mid d_r$. So we know

$$M \cong \frac{D^m}{\langle (d_1, 0, \cdots, 0), (0, d_2, 0, \cdots, 0), \cdots, (0, \cdots, 0, d_r, 0, \cdots, 0) \rangle}.$$

This is just

$$\frac{D}{\langle d_1 \rangle} \oplus \frac{D}{\langle d_2 \rangle} \oplus \cdots \oplus \frac{D}{\langle d_r \rangle} \oplus D \oplus \cdots \oplus D,$$

with $m - r$ copies of D. If $d_i \in U(D)$ we see that $R/\langle d_i \rangle = 0$. So we can delete this unit d_i. $\square$

A different form of the classification of finitely-generated modules over a Euclidean domain can be found in Exercise (20). We point out that all results so far in this section hold also for principal ideal domains, see [J, Chapter 3].

We can consider any additive abelian group as a $\mathbb{Z}$-module. Taking $D = \mathbb{Z}$, and applying the classification of finitely generated D-modules, we recover the following classification theorem for finitely-generated abelian groups (Theorem 1.6.4).

Theorem 3.3.13 (Classification of finitely-generated abelian groups). *Every nontrivial finitely generated abelian group G is isomorphic to a group of direct product of nontrivial cyclic groups*

$$\mathbb{Z}_{m_1} \times \mathbb{Z}_{m_2} \times \cdots \times \mathbb{Z}_{m_r} \times \mathbb{Z}^n,$$

where $r, n \in \mathbb{Z}^+, m_1, \cdots, m_r \in 1 + \mathbb{Z}^+$ with $m_i | m_{i+1}$ for $i = 1, 2, \cdots, r - 1$.

Example 3.3.1. *Find the Smith normal form of the integer matrix*

$$A = \begin{bmatrix} 0 & 1 & 1 \\ 1 & 0 & -3 \\ 1 & -3 & 0 \end{bmatrix}.$$

Proof. We compute that

$$A = \begin{bmatrix} 0 & 1 & 1 \\ 1 & 0 & -3 \\ 1 & -3 & 0 \end{bmatrix} \xrightarrow[C_3 - C_2 \to C_3]{R_3 - R_2 \to R_3} \begin{bmatrix} 0 & 1 & 0 \\ 1 & 0 & -3 \\ 0 & -3 & 6 \end{bmatrix}$$

$$\xrightarrow[R_3 + 3R_1 \to R_3]{C_3 + 3C_1 \to C_3} \begin{bmatrix} 0 & 1 & 0 \\ 1 & 0 & 0 \\ 0 & 0 & 6 \end{bmatrix} \xrightarrow{R_2 + R_1 \to R_1} \begin{bmatrix} 1 & 1 & 0 \\ 1 & 0 & 0 \\ 0 & 0 & 6 \end{bmatrix}$$

$$\xrightarrow[C_2 - C_1 \to C_2]{R_2 - R_1 \to R_2} \begin{bmatrix} 1 & 0 & 0 \\ 0 & -1 & 0 \\ 0 & 0 & 6 \end{bmatrix} \xrightarrow{-R_2 \to R_2} \begin{bmatrix} 1 & 0 & 0 \\ 0 & 1 & 0 \\ 0 & 0 & 6 \end{bmatrix},$$

which is the Smith normal form of A. $\square$

3.4. Noetherian rings.

We now introduce Noetherian rings, left Noetherian rings, and right Noetherian rings in this section and prove Hilbert basis theorem.

Definition 3.4.1 (Noetherian ring). *A ring R is **Noetherian** (or **left Noetherian**, or **right Noetherian**, respectively) if any chain of ideals (or left ideals, or right ideals, respectively) of R*

$$I_1 \subseteq I_2 \subseteq I_3 \subseteq \cdots ,$$

satisfies ACC, i.e., there is some $n \in \mathbb{N}$ such that $I_n = I_{n+1} = I_{n+2} = \cdots$.

Example 3.4.1. (a). *Every finite ring is Noetherian, left Noetherian and right Noetherian.*
  (b). *Every principal ideal domain D is Noetherian. This is because any nonzero and nonunit $a \in D$ has only finitely factors up to associates.*
  (c). *The ring $\mathbb{Z}[x_1, x_2, x_3, \cdots]$ is not Noetherian since it has the chain of strictly increasing ideals*

$$\langle x_1 \rangle \subset \langle x_1, x_2 \rangle \subset \langle x_1, x_2, x_3 \rangle \subset \cdots .$$

We have the following proposition that makes Noetherian rings much more concrete, and makes it obvious why PIDs are Noetherian.

Definition 3.4.2 (Finitely generated ideal). *An ideal I of a ring R is **finitely generated** if there are $r_1, \cdots, r_n \in R$ such that $I = \langle r_1, \cdots, r_n \rangle$.*

Similarly, we can define finitely-generated left ideals and finitely-generated right ideals.

Proposition 3.4.3. *A ring R is Noetherian (or left Noetherian, or right Noetherian, respectively) if and only if every ideal (or left ideal, or right ideal, respectively) of R is finitely generated.*

Proof. ($\Rightarrow$). Suppose every ideal of R is finitely generated. Given the chain $I_1 \subseteq I_2 \subseteq \cdots$, we have the ideal

$$I = I_1 \cup I_2 \cup I_3 \cup \cdots .$$

We know I is finitely generated, say $I = \langle r_1, \cdots, r_n \rangle$, with $r_i \in I_{k_i}$. Let

$$n = \max_{i=1,\cdots,n} \{k_i\}.$$

Then $r_1, \cdots, r_n \in I_K$. So $I_n = I$, and furthermore $I_n = I_{n+1} = I_{n+2} = \cdots$.

($\Leftarrow$). Suppose there is an ideal $I \lhd R$ that is not finitely generated. We pick $r_1 \in I$. Since I is not finitely generated, we know $\langle r_1 \rangle \neq I$. So we can find some $r_2 \in I \setminus \langle r_1 \rangle$.

Again $\langle r_1, r_2 \rangle \neq I$. So we can find $r_3 \in I \setminus \langle r_1, r_2 \rangle$. We continue on, and then can find an infinite strictly ascending chain

$$\langle r_1 \rangle \subseteq \langle r_1, r_2 \rangle \subseteq \langle r_1, r_2, r_3 \rangle \subseteq \cdots.$$

So R is not Noetherian.

For left Noetherian, or right Noetherian cases the proof is similar. $\square$

If R is Noetherian, not necessarily every subring of R has to be Noetherian. For example, since $\mathbb{Z}[x_1, x_2, \cdots]$ is an integral domain, we can take its field F of fractions, which is a field, hence Noetherian, but $\mathbb{Z}[x_1, x_2, \cdots]$ is a subring of F. For quotient rings we have the following result.

Proposition 3.4.4. *Let R be a Noetherian ring and $I \unlhd R$. Then R/I is Noetherian.*

Proof. Consider the natural homomorphism

$$\pi : R \to R/I, \quad x \mapsto x + I.$$

Let $J \lhd R/I$. We want to show that J is finitely generated. We know that $\pi^{-1}(J) \unlhd R$, and is hence finitely generated, since R is Noetherian. So $\pi^{-1}(J) = \langle r_1, \cdots, r_n \rangle$ for some $r_1, \cdots, r_n \in R$. Then J is generated by $\pi(r_1), \cdots, \pi(r_n)$. So R/I is Noetherian by Proposition 3.4.3. $\square$

Now we can prove the following powerful theorem which was, surprisingly, proven by David Hilbert (1862–1943) in 1890:

Theorem 3.4.5 (Hilbert basis theorem). *Let R be a Noetherian ring. Then so is $R[x]$.*

Proof. To the contrary, suppose $\mathfrak{a} \subseteq R[x]$ is a non-finitely-generated ideal. Then by recursion there is a sequence $\{f_0, f_1, \ldots\} \subset \mathfrak{a}$ such that if $\mathfrak{b}_n$ with $n \geq 1$ is the ideal generated by $f_0, \ldots, f_{n-1}$, then $f_n \in \mathfrak{a} \setminus \mathfrak{b}_n$ is of minimal degree. It is clear that $\{\deg(f_0), \deg(f_1), \ldots\}$ is a non-decreasing sequence of nonnegative integers. Let a_n be the leading coefficient of f_n and let $\mathfrak{b}$ be the ideal of R generated by

$a_0, a_1, \ldots$. Since R is Noetherian the chain of ideals

$$\langle a_0 \rangle \subset \langle a_0, a_1 \rangle \subset \langle a_0, a_1, a_2 \rangle \subset \cdots$$

must terminate. Suppose that $\mathfrak{b} = \langle a_0, \ldots, a_{n-1} \rangle$ for some integer n. So in particular,

$$a_n = \sum_{i<n} \sum_{j=1}^{n_i} u_{i,j} a_i v_{i,j}, \qquad u_{i,j}, v_{i,j} \in R,$$

where the sum is finite. Now consider

$$g = \sum_{i<n} \sum_{j=1}^{n_i} u_{i,j} x^{\deg(f_n)-\deg(f_i)} f_i v_{i,j} \in \mathfrak{b}_n,$$

whose leading term is equal to that of f_n. However, $f_n \notin \mathfrak{b}_n$, which means that

$$f_n - g \in \mathfrak{a} \setminus \mathfrak{b}_n$$

has degree less than f_n, contradicting the minimality.   □

Modifying the above proof we can have the following result:

Theorem 3.4.6. *Let R be a left (or right) Noetherian ring. Then so is $R[x]$.*

A direct consequence of the above results are the following corollary.

Corollary 3.4.7. *If R is a Noetherian ring (or left Noetherian ring, or right Noetherian ring, respectively), then the polynomial ring $R[x_1, \ldots, x_n]$ in commutative indeterminates $x_1, \ldots, x_n$ is a Noetherian ring (or left Noetherian ring, or right Noetherian ring, respectively).*

We now explain an application of the above theorem. Let $S \subseteq F[x_1, x_2, \cdots, x_n]$ be any set of polynomials where F is an arbitrary field. We define the zero-locus $Z(S)$ to be the set of points in F^n on which the functions in S simultaneously vanish, that is

$$Z(S) = \{\alpha \in F^n \mid f(\alpha) = 0 \text{ for all } f \in S\}.$$

A subset V of F^n is called an **affine algebraic set** if $V = Z(S)$ for some $S \subseteq F[x_1, x_2, \cdots, x_n]$.

We view this as a set of equations $f = 0$ for each $f \in S$. The claim is that to solve the potentially infinite set of equations S, we actually only have to solve finitely many equations.

Consider the ideal $\langle S \rangle \trianglelefteq F[x_1, \cdots, x_n]$. By the Hilbert basis theorem, there is a finite list $f_1, \cdots, f_k$ such that

$$\langle f_1, \cdots, f_k \rangle = \langle S \rangle.$$

We can easily see that

$$Z(S) = Z(\langle S \rangle) = Z(f_1, \cdots, f_k).$$

So solving S is the same as solving $f_1, \cdots, f_k$. This is extremely useful.

Next we will only introduce some important concepts in Commutative Algebra.

Definition 3.4.8. *The* **Krull dimension** *of a unital ring R, denoted by $\dim R$, is the maximum length n of a chain $I_0 \subset I_1 \subset \cdots \subset I_n$ of prime ideals of R. If there is no upper bound on the length of such a chain, we take $n = \infty$.*

Example 3.4.2.   *(a). A field F has Krull dimension 0.*
   (b). The polynomial ring $F[x_1, x_2, \cdots, x_n]$ has Krull dimension n.
   (c). A principal ideal domain that is not a field has Krull dimension 1.
   (d). The Krull dimension of the non-Noetherian ring $F[x_1, x_2, \cdots]$ is infinity, where F is a field. We have the infinite chain of prime ideals

$$\langle x_1 \rangle \subset \langle x_1, x_2 \rangle \subset \langle x_1, x_2, x_3 \rangle \subset \cdots .$$

Definition 3.4.9. *The* **height of a prime ideal** *P of a unital ring R is the maximum length n of a chain of prime ideals $I_0 \subset I_1 \subset \cdots \subset I_n = P$ of R.*

Definition 3.4.10. *Let P be a maximal ideal of a unital ring R. The sequence $a_1, \cdots, a_t$ of nonzero elements in P is a* **regular sequence** *for R, if each a_i is not a zero-divisor of $R/\langle a_1, \cdots, a_{i-1} \rangle$.*

Definition 3.4.11. *The* **depth of a ring** *R is the maximal number of elements in some regular sequence, denoted by $\mathrm{depth}(R)$. A ring in which $\mathrm{depth}(R) = \dim(R)$ is called a* **Cohen–Macaulay ring**.

Example 3.4.3. *Let F be a field.*
   (a). The rings $F[x_1, \cdots, x_n]$, $F[x, y]/\langle xy \rangle$, $F[x, y, z]/\langle xy, xz, yz \rangle$, $F[x, y, z]/\langle xy - z \rangle$, and $F[x, y, z, w]/\langle xy - zw \rangle$ are all Cohen–Macaulay.

(b). None of the rings $F[x,y]/\langle x^2, xy \rangle, F[x,y,z]/\langle xy, xz \rangle$, or $F[x,y,z,w]/\langle wy, wz, xy, xz \rangle$ is Cohen–Macaulay.

We leave the proofs as an exercise.

3.5. Exercises.

(1) Classify all simple modules over the ring $R = (\mathbb{Z}, +, \cdot)$.
(2) Classify all simple modules over the ring $R = (2\mathbb{Z}, +, \cdot)$.
(3) Classify all simple modules over the ring $R = (\mathbb{Q}[x], +, \cdot)$.
(4) Classify all simple modules over the ring $R = (\mathbb{R}[x], +, \cdot)$.
(5) Find the Smith normal form of the integer matrix

$$A = \begin{bmatrix} 5 & -417 & 129 & 50 \\ -6 & 111 & -36 & 6 \\ 5 & -672 & 210 & 74 \\ -7 & 255 & -81 & -10 \end{bmatrix}.$$

(6) Find the invariant factors of the integer matrix

$$A = \begin{bmatrix} 2 & 4 & 4 \\ -6 & 6 & 12 \\ 10 & 4 & 16 \end{bmatrix}.$$

(7) Without doing any elementary row or column operations find the Smith normal form of the integer matrix

$$A = \begin{bmatrix} 2 & 0 & 0 \\ 0 & 3 & 0 \\ 0 & 0 & 4 \end{bmatrix}, \quad B = \begin{bmatrix} 15 & 0 & 0 \\ 0 & 10 & 0 \\ 0 & 0 & 6 \end{bmatrix}.$$

(8) Find the Smith normal form of the matrix over $\mathbb{C}[x]$:

$$\begin{bmatrix} 1-x & 1 & 1 \\ 0 & 1-x & 0 \\ 0 & 1 & 2-x \end{bmatrix}.$$

(9) Find the invariant factors of the matrix over $\mathbb{Q}[x]$:

$$\begin{bmatrix} 2 & 2x+3 & 2x^2+3x \\ 1 & 6x & 6x^2+6x \\ 1 & 3 & x \end{bmatrix}.$$

(10) Suppose that V be a finitely generated module over $\mathbb{C}[x]$ that is not a free module over $\mathbb{C}[x]$. Show that x has an eigenvector on V.

(11) Find a basis for the $\mathbb{Z}[x]$-submodule of $\mathbb{Z}[x]$-module $\mathbb{Z}[x]^3$ generated by $u_1 = (3x+1, 3x+2, (x+1)^2), u_1 = (x+1, 2x+1, x^2), u_1 = (1+x, 1, 2x+1)$.

(12) Let M be a left R-module. Show that M is finitely generated if there exists a submodule $N \subset M$ such that N and M/N are both finitely generated.

(13) Show that if $x^2 = 0$ implies $x = 0$, for all x in the ring R, then all idempotent elements of R are central.

(14) Let R be a commutative ring with a unique maximal ideal I, and let M be a nonzero finitely generated R-module. Show that $\mathrm{Hom}_R(M, R/I) \neq 0$.

(15) In Definition 3.2.2 show that the R-module homomorphism $\tilde{\varphi}: M \to N$ is uniquely determined by the map φ.

(16) Let R be a unital ring, and let M be a left R-module that has a minimal submodule S such that $M/S = S$. Prove that either S is a direct summand of M, in which case $M = S \oplus S$, or else S is the only proper nontrivial submodule of M.

(17) Let A and B be finitely generated modules over a Euclidean domain D. If $A \oplus A \cong B \oplus B$, prove that $A \cong B$.

(18) Let R be any ring, A a R-module, and $A = B \oplus C = D \oplus E$ two direct sum decompositions of A. Let f be the projection on B, restricted to D, let g be the projection on E, restricted to C.

(a). If f is one-to-one, so is g.

(b). If f is onto, so is g.

(19) Let D be a Euclidean domain, and $a, b \in D$ be such that $\gcd(a, b) = 1$. Show that $D/\langle ab \rangle \cong D/\langle a \rangle \oplus D/\langle b \rangle$ as D-modules.

(20) Let D be a Euclidean domain, and M be a finitely-generated R-module. Show that

$$M \cong N_1 \oplus N_2 \oplus \cdots \oplus N_r,$$

where each N_i is either D or is $D/\langle p^n \rangle$ for some prime $p \in D$ and some $n \geq 1$.

(21) The ideal $I = \langle 2, x \rangle$ of the ring $R = (\mathbb{Z}[x], +, \cdot)$ is not a direct sum of cyclic $\mathbb{Z}[x]$-modules.

(22) Determine all simple modules over the ring $R = (\mathbb{Z}[x], +, \cdot)$.

(23) In the non-Noetherian ring $F[x_1, x_2, \cdots]$, where F is a field, show that the ideals $\langle x_1, x_2, \cdots, x_n \rangle$ are prime for any positive integer n.

(24) Show that an R-module homomorphism between two simple modules over a unital ring R is zero or an isomorphism.

(25) Let M be a nonzero finitely generated module over a ring R. Show that M has a maximal submodule.

(26) Let R be a unital ring, and $n \in \mathbb{N}$. Show that R is Noetherian if and only if $M_n(R)$ is Noetherian.

(27) Prove Theorem 3.4.6.

(28) Let R be a left-Noetherian ring and M be a finitely-generated left R-module. Show that any submodule of M is finitely-generated. (Hints: Use induction on the size of the generating set.)

(29) A commutative ring R is finitely generated if there are $u_1, \cdots,$ $u_n \in R$ such that every element of R is a finite sum of elements of the form

$$au_1^{k_1} \cdots u_n^{k_n}, \text{ where } a \in \mathbb{Z}, k_i \in \mathbb{N}.$$

Show that every finitely generated commutative ring R is Noetherian.

(30) Let M be a finitely generated module over a unital commutative ring R, and $I \leq R$ such that $IM = M$. Show that there exists $a \in I$ such that $(1 + a)M = 0$. (Hints: Use adjoint matrix.)

(31) (Cohen's Theorem). Let R be a unital commutative ring. Show that R is Noetherian if and only if all prime ideals of R are finitely generated. (Hints: By contradiction.)

(32) Prove Example 3.4.3.

(33) Show that the ring $\mathbb{C}[x_1, \cdots, x_5]/I$ where $I = \langle x_1 x_5, x_1 x_2, x_2 x_3, x_3 x_4, x_4 x_5 \rangle$ is Cohen–Macaulay.

4. Fields and Extension Fields

Historically, three algebraic disciplines led to the concept of a field: the question of solving polynomial equations, algebraic number theory, and algebraic geometry. A first step towards the notion of a field was made in 1770 by Joseph-Louis Lagrange. The first clear definition of an abstract field (1893) is due to Heinrich Martin Weber. In particular, Weber's notion included the field $\mathbb{Z}_p$.

In this chapter we will study the basic theory on the extension fields, including algebraic extensions and transcendental extensions, and establish basic results for finite fields.

4.1. Prime fields and extension fields.

We first introduce some basic definitions on prime fields and extension fields.

Let E be a field. Recall that a subset F of E is called a subfield of E if F is a field with the same operations as E, denoted by $F \leq E$. It is easy to see that if $F \subseteq E$, then F is a subfield of E if and only if

$$1, a - b, ab^{-1} \in F, \forall a, b \in F \text{ with } b \neq 0.$$

It is clear that the intersection of all subfields of E is again a subfield. The following theorem describes the structure of the smallest subfield of E.

Theorem 4.1.1. *Let E be a field and P the smallest subfield of E.*

(1). If $\mathrm{char}(E) = 0$, *then P is isomorphic to $\mathbb{Q}$.*

(2). If $\mathrm{char}(E) = p$ *is a prime, then P is isomorphic to $\mathbb{Z}_p$.*

Proof. Let F be any subfield of E. Then $1 \in F$, and $\mathbb{Z} \cdot 1 \subseteq F$.

(1). If $\mathrm{char}(E) = 0$, then the subring $\mathbb{Z} \cdot 1$ of F is isomorphic to $\mathbb{Z}$. So F must contain a quotient field of this subring and that this quotient field must be isomorphic to $\mathbb{Q}$. Thus, F contains a subfield isomorphic to $\mathbb{Q}$. Therefore, $P \cong \mathbb{Q}$.

(2). If $\mathrm{char}(E) = p$, then the subring $\mathbb{Z} \cdot 1$ of F is isomorphic to $\mathbb{Z}_p$. It follows that $P \cong \mathbb{Z}_p$. $\qquad\square$

Definition 4.1.2.   *(1). The fields $\mathbb{Q}$ and $\mathbb{Z}_p$ for any prime p are call* **prime fields**.

(2). If $F \leq E$ are fields, then E is called an **extension field of F**.

We see that every field F is an extension field of the prime subfield P of F. In particular, $\mathbb{R}$ is an extension field of $\mathbb{Q}$, and $\mathbb{C}$ is an extension field of both $\mathbb{R}$ and $\mathbb{Q}$. If $F \leq E$ are fields, we can consider E as a vector space over F. We denote $[E : F] = \dim_F E$, the dimension of E as a vector space over F.

Definition 4.1.3. *Let E be an extension of a field F. If $[E : F] = n < \infty$, then E is called a* **finite extension of degree n over F**. *Otherwise, we say that E is an infinite extension field over F.*

Although a field E is a finite extension of a field F, it does not mean that E is a finite field. It means that E is a finite-dimensional vector space over F.

Notice that $[E : F] = 1$ if and only if $E = F$.

Theorem 4.1.4. *Suppose $F \leq K \leq E$ are fields.*

(1). $[E : F]$ is finite if and only if $[E : K]$ and $[K : F]$ are finite.
(2). If $[E : F] < \infty$, then $[E : F] = [E : K][K : F]$.

Proof. (1). ($\Rightarrow$). Assume that $[E : F] < \infty$. It follows that $[K : F] < \infty$ since K is a subspace of E over F. Let $\gamma = \{\gamma_1, \gamma_2, \cdots, \gamma_n\}$ be a basis for E as a vector space over F. Then every element of E is a linear combination of γ with coefficients in F and hence with coefficients in K. Therefore, by a result from linear algebra, a subset of γ forms a basis of E over K. Thus $[E : K] < \infty$.

($\Leftarrow$). Suppose that $[E : K]$ and $[K : F]$ are finite. Let $\alpha = \{\alpha_1, \alpha_2, \cdots, \alpha_r\}$ be a basis of K as a vector space over F, and let $\beta = \{\beta_1, \beta_2, \cdots, \beta_s\}$ be a base of E as a vector space over K. It is enough to show that the rs elements $\alpha_i \beta_j$ form a basis for E as a vector space over F.

For any $\gamma \in E$ since β is a basis for E over K, we have

$$\gamma = \sum_{j=1}^{s} b_j \beta_j$$

where $b_j \in K$. Since α is a basis for K over F, we have

$$b_j = \sum_{i=1}^{r} a_{ij}\alpha_i$$

where $a_{ij} \in F$. Then we have

$$\gamma = \sum_{j=1}^{r} \left(\sum_{i=1}^{s} a_{ij}\alpha_i \right) \beta_j = \sum_{i,j} a_{ij}(\alpha_i\beta_j),$$

so every element of E is an F-linear combination of the rs vectors $\alpha_i\beta_j$.

Next we show that the rs elements $\alpha_i\beta_j$ are linearly independent over F. Suppose that $\sum_{i,j} a_{ij}(\alpha_i\beta_j) = 0$, with $a_{ij} \in F$. Then

$$\sum_{j=1}^{s} \left(\sum_{i=1}^{r} a_{ij}\alpha_i \right) \beta_j = 0$$

and $\sum_{i=1}^{n} a_{ij}\alpha_i \in K$. Since β is independent over E, we see that

$$\sum_{i=1}^{r} a_{ij}\alpha_i = 0$$

for all j. Since α is independent over F, so $a_{ij} = 0$ for all i and j. Thus $\{\alpha_i\beta_j : i = 1, 2, \cdots, r; j = 1, 2, \cdots, s\}$ forms a basis for E over F and

$$[E : F] = [E : K][K : F] < \infty,$$

and (2) follows also.   □

A direct consequence of the above theorem is the following result.

Corollary 4.1.5. *Let $E_1 \leq E_2 \leq \cdots \leq E_s$ be fields and $[E_{i+1} : E_i] < \infty$ for $i = 1, \cdots, s - 1$. Then $[E_s : E_1]$ is finite and*

$$[E_s : E_1] = [E_s : E_{s-1}][E_{s-1} : E_{s-2}] \cdots [E_2 : E_1].$$

A very useful observation is that if $[E : F]$ is finite, then $[K : F] \mid [E : F]$ for any $K \leq E$.

Definition 4.1.6. *Let $F \leq E$ be fields and $\emptyset \neq S \subseteq E$. The field*

$$F(S) = \bigcap_{S \cup F \subseteq K \leq E} K$$

is called the **subfield generated by** S **over** F.

It is not hard to see that the subfield $F(S)$ is the minimum subfield of E containing F and S. If $S = \{a_1, a_2, \cdots, a_n\}$ is a finite subset of E, we write $F(S)$ as $F(a_1, a_2, \cdots, a_n)$. Clearly, elements in $F(a_1, a_2, \cdots, a_n)$ are of the form:

$$\frac{f(a_1, a_2, \cdots, a_n)}{g(a_1, a_2, \cdots, a_n)},$$

where $f, g \in F[x_1, x_2, \cdots, x_n]$ with $g(a_1, a_2, \cdots, a_n) \neq 0$.

In particular, for any $a \in E$,

$$F(a) = \left\{ \frac{f(a)}{g(a)} : f, g \in F[x], g(a) \neq 0 \right\}.$$

Definition 4.1.7. *If $F \leq E$ be fields and $E = F(a)$ for some $a \in E$, then E is called a* **simple extension field** *of F.*

Example 4.1.1. *We regard $\mathbb{R}$ as an extension field of $\mathbb{Q}$. It is easy to see that $\mathbb{Q}(\sqrt{2}) = \{a + b\sqrt{2} : a, b \in \mathbb{Q}\} = \mathbb{Q}[\sqrt{2}]$.*

4.2. Algebraic and transcendental elements.

In this section we always assume that F is a field. We firstly prove an important result that follows quickly and elegantly. This theorem is named after Leopold Kronecker (1823–1891) who proved it in 1884.

Theorem 4.2.1 (Kronecker's Theorem). *For any field F and $f(x) \in F[x]$ with $\deg(f(x)) > 0$, there exists an extension field E of F and an $\alpha \in E$ such that $f(\alpha) = 0$.*

Proof. We have known that $f(x)$ can be written as a product of irreducible polynomials in $F[x]$. Now let $p(x)$ be an irreducible factor of $f(x)$. Thus it is sufficient to find an extension field E of F containing an element α such that $p(\alpha) = 0$.

Since $p(x)$ is irreducible in $F[x]$, from Theorem 1.5.13 we see that $\langle p(x) \rangle$ is a maximal ideal of $F[x]$, and further $E = F[x]/\langle p(x) \rangle$ is a field by Theorem 1.5.4. We first want to identify F with a subfield of $F[x]/\langle p(x) \rangle$ in a natural way. Define the map

$$\psi : F \to E, \quad \psi(a) = a + \langle p(x) \rangle, \forall a \in F.$$

It is easy to see that ψ is a ring homomorphism. If $\psi(a) = \psi(b)$ for $a, b \in F$, that is, if $a + \langle p(x) \rangle = b + \langle p(x) \rangle$ for some $a, b \in F$, then $a - b \in \langle p(x) \rangle$, i.e., $p(x)|a - b$. Thus we deduce that $a - b = 0$, so $a = b$. Then ψ is one to one.

So ψ maps F one-to-one onto a subfield of $F[x]/\langle p(x) \rangle$. We may identify F with $\text{Im}(\psi) = \{a + \langle p(x) \rangle : a \in F\}$. Thus we shall view $E = F[x]/\langle p(x) \rangle$ as an extension field of F. Next we show that E contains a zero of $p(x)$.

Consider $\alpha = x + \langle p(x) \rangle \in E$. Take the **evaluation homomorphism**

$$\phi_\alpha : F[x] \to E, \quad \phi_\alpha(g(x)) = g(\alpha), \forall g(x) \in F[x].$$

If $p(x) = a_0 + a_1 x + \cdots + a_n x^n$, where $a_i \in F$, then we have

$$\phi_\alpha(p(x)) = a_0 + a_1(x + \langle p(x) \rangle) + \cdots + a_n(x + \langle p(x) \rangle)^n$$

in $E = F[x]/\langle p(x) \rangle$. Therefore,

$$p(\alpha) = a_0 + a_1 x + \cdots + a_n x^n + \langle p(x) \rangle = p(x) + \langle p(x) \rangle = \langle p(x) \rangle = 0$$

in $F[x]/\langle p(x) \rangle$. Thus, $p(\alpha) = 0$, and therefore $f(\alpha) = 0$. $\qquad \square$

We illustrate the construction involved in the proof by an example.

Example 4.2.1. *Take $F = \mathbb{R}$ and let $f(x) = x^2 + 1 \in \mathbb{R}[x]$. We know that $f(x)$ has no zeros in $\mathbb{R}$ and thus is irreducible in $\mathbb{R}[x]$. Then $\langle x^2 + 1 \rangle$ is a maximal ideal in $\mathbb{R}[x]$, and further $\mathbb{R}[x]/\langle x^2 + 1 \rangle$ is a field. Identifying $r \in \mathbb{R}$ with $r + \langle x^2 + 1 \rangle$ in $\mathbb{R}[x]/\langle x^2 + 1 \rangle$, we can view $\mathbb{R}$ as a subfield of $E = \mathbb{R}[x]/\langle x^2 + 1 \rangle$. Take $\alpha = x + \langle x^2 + 1 \rangle \in \mathbb{R}[x]/\langle x^2 + 1 \rangle$. In $\mathbb{R}[x]/\langle x^2 + 1 \rangle$, we compute*

$$\alpha^2 + 1 = (x + \langle x^2 + 1 \rangle)^2 + (1 + \langle x^2 + 1 \rangle) = (x^2 + 1) + \langle x^2 + 1 \rangle = 0.$$

So α is a zero of $x^2 + 1$. We shall identify $\mathbb{R}[x]/\langle x^2 + 1 \rangle$ with $\mathbb{C}$.

In the following we put an element of an extension field E of a field F into one of two categories.

Definition 4.2.2. *Let $F \leq E$ be fields and $\alpha \in E$. If $f(\alpha) = 0$ for some nonzero $f(x) \in F[x]$, then α is called **algebraic over F**. Otherwise α is called **transcendental over F**.*

Example 4.2.2. (1). *Regard $\mathbb{C}$ as an extension field of $\mathbb{Q}$. Since $\sqrt{3}$ is a zero of $x^2 - 3 \in \mathbb{Q}[x]$, we see that $\sqrt{3}$ is an algebraic element over $\mathbb{Q}$. Also, i is an algebraic element over $\mathbb{Q}$, since i is a zero of $x^2 + 1 \in \mathbb{Q}[x]$.*

   (2). *It is well known (but the proof is not easy) that the real numbers π and e are transcendental over $\mathbb{Q}$. Here e is the base for the natural logarithm.*

(3). The real number π is transcendental over $\mathbb{Q}$. However, π is algebraic over $\mathbb{R}$, for it is a zero of $x - \pi \in \mathbb{R}[x]$.

(4). Let $\alpha = \sqrt{1 + \sqrt{2}} \in \mathbb{R}$. It is easy to see that α is algebraic over $\mathbb{Q}$. Since $\alpha^2 - 1 = \sqrt{2}$ and $(\alpha^2 - 1)^2 = 2$. Therefore $\alpha^4 - 2\alpha^2 - 1 = 0$, so α is a zero of the polynomial $f(x) = x^4 - 2x^2 - 1 \in \mathbb{Q}[x]$.

Definition 4.2.3. *Regard $\mathbb{C}$ as an extension field of $\mathbb{Q}$. The complex number α is called an* **algebraic number** *if it is algebraic over $\mathbb{Q}$, otherwise, α is called a* **transcendental number**.

The definition above connect these ideas in the theory of field with those in number theory. There is an extensive and elegant theory of algebraic numbers in number theory.

The next theorem provides a useful characterization of algebraic and transcendental elements over F in an extension field E of F.

Theorem 4.2.4. *Let $F \leq E$ be fields and let $\alpha \in E$. Then α is transcendental over F if and only if the evaluation homomorphism $\phi_\alpha : F[x] \to E$ defined by $\phi_\alpha(g(x)) = g(\alpha)$ for $g(x) \in F[x]$ is one-to-one. Thus, $F[x] \cong \mathrm{im}(\phi_\alpha) = F[\alpha]$.*

Proof. By definition, α is transcendental over F if and only if $f(\alpha) \neq 0$ for all nonzero $f(x) \in F[x]$, if and only if $\phi_\alpha(f(x)) \neq 0$ for all nonzero $f(x) \in F[x]$, if and only if the kernel of ϕ_α is 0, if and only if ϕ_α is one-to-one. The isomorphism $F[x] \cong \mathrm{im}(\phi_\alpha) = F[\alpha]$ follows from Theorem 1.2.10.   □

The next theorem plays a central role in our later sections.

Theorem 4.2.5. *Let $F \leq E$ be fields and let $\alpha \in E$ be algebraic over F.*

(1). There is an unique monic irreducible polynomial $p(x) \in F[x]$ such that $p(\alpha) = 0$.

(2). If $f(\alpha) = 0$ for $f(x) \in F[x]$, then $p(x)|f(x)$.

Proof. (1). Let $\phi_\alpha : F[x] \to E$ be the evaluation homomorphism. We know that $\ker(\phi_\alpha) \trianglelefteq F[x]$. Since $F[x]$ is a PID, then $\ker(\phi_\alpha) = \langle p(x) \rangle$ for some monic $p(x) \in F[x]$.

Now we prove that $p(x)$ is irreducible. Suppose that $p(x) = r(x)s(x)$ with $\deg(r(x)), \deg(s(x)) < \deg(p(x))$. Since $p(\alpha) = 0$ then $r(\alpha)s(\alpha) = 0$, yielding that $r(\alpha) = 0$ or $s(\alpha) = 0$, since E is a field.

Say $r(\alpha) = 0$. Then $r(x) \in \ker(\phi_\alpha) = \langle p(x) \rangle$. This is impossible since $0 < \deg(r(x)) < \deg(p(x))$. Hence $p(x)$ is irreducible over F.

If $q(x) \in F[x]$ is monic irreducible such that $q(\alpha) = 0$. Then $q(x) \in \ker(\phi_\alpha) = \langle p(x) \rangle$. Thus $p(x)|q(x)$. So $p(x) = q(x)$. Part (1) follows.

(2). Now the principal ideal $\langle p(x) \rangle$ consists precisely of those polynomials of $F[x]$ having α as a zero. If $f(\alpha) = 0$ for $f(x) \in F[x]$, then $f(x) \in \ker(\phi_\alpha) = \langle p(x) \rangle$. It follows that $p(x)|f(x)$. $\qquad\square$

Definition 4.2.6. *Let $F \leq E$ be fields and $\alpha \in E$ be algebraic over F. The unique monic polynomial $p(x)$ with $p(\alpha) = 0$ is called the* **irreducible polynomial** *of α over F and will be denoted by* $\mathrm{irr}(\alpha, F)$. *The degree of* $\mathrm{irr}(\alpha, F)$ *is the* **degree** *of α over F, denoted by* $\deg(\alpha, F)$.

The irreducible polynomial of α over F is also called minimal polynomial of α over F in some textbooks.

Example 4.2.3. *By the previous examples, we know that*

(1). $\mathrm{irr}(\sqrt{2}, \mathbb{Q}) = x^2 - 2$; $\mathrm{irr}(\sqrt{2}, \mathbb{R}) = x - \sqrt{2}$.

(2). $\mathrm{irr}(\sqrt{-1}, \mathbb{Q}) = x^2 + 1$.

(3). We see that for $\alpha = \sqrt{1 + \sqrt{3}}$ in $\mathbb{R}$, α is a zero of $x^4 - 2x^2 - 2 \in \mathbb{Q}[x]$. Since $x^4 - 2x^2 - 2$ is irreducible over $\mathbb{Q}$ (by Schönemann-Eisenstein Criterion with $p = 2$), we know that $\mathrm{irr}(\sqrt{1 + \sqrt{3}}, \mathbb{Q}) = x^4 - 2x^2 - 2$. Thus $\sqrt{1 + \sqrt{3}}$ is algebraic of degree 4 over $\mathbb{Q}$.

Let $F \leq E$ be fields and $\alpha \in E$. Let $\phi_\alpha : F[x] \to E$ be the evaluation homomorphism. By the previous results we should consider the following two cases.

Case 1: α is algebraic over F. Then $\ker(\phi_\alpha) = \langle \mathrm{irr}(\alpha, F) \rangle$ which is a maximal ideal of $F[x]$. Therefore, $F[x]/\langle \mathrm{irr}(\alpha, F) \rangle$ is a field and is isomorphic to the image $\phi_\alpha(F[x])$ in E. This subfield $\phi_\alpha(F[x])$ of E is then the smallest subfield of E containing F and α. Thus, in this case we have $\phi_\alpha(F[x]) = F[\alpha] = F(\alpha)$.

Case 2: α is transcendental over F. Then ϕ_α gives an isomorphism of $F[x]$ with a subdomain $F[\alpha]$ of E. Thus in this case $\phi_\alpha(F[x])$ is not a field but an integral domain. We see that E contains a field of quotients of $F[\alpha]$, which is just the smallest subfield $F(\alpha)$ of E containing F and α.

Example 4.2.4. *Since e is transcendental over $\mathbb{Q}$, the field $\mathbb{Q}(e)$ is isomorphic to the field $\mathbb{Q}(x)$ of rational functions over $\mathbb{Q}$ in the indeterminate x.*

The next theorem characterize algebraic elements over a field F.

Theorem 4.2.7. *Let $E = F(\alpha)$ be an extension of F where $\alpha \in E$ is algebraic over F. Let $\deg(\alpha, F) = n \geq 1$. As a vector space over F, then E has a basis $\{1, \alpha, \alpha^2, \cdots, \alpha^{n-1}\}$. Consequently $\deg(\alpha, F) = [F(\alpha) : F]$.*

Proof. Since $\alpha \in E$ is algebraic over F, we know that $E = F(\alpha) = F[\alpha]$. Assume that

$$\mathrm{irr}(\alpha, F) = p(x) = x^n + a_{n-1}x^{n-1} + \cdots + a_0$$

be the irreducible polynomial of α over F. For any element $\beta \in E$, we can write $\beta = f(\alpha)$ for some polynomial $f(x) \in F[x]$. By the Euclidean Algorithm, write $f(x) = p(x)q(x) + r(x)$, where $q(x), r(x) \in F[x]$ and $r(x) = 0$ or $\deg(r(x)) < \deg(p(x)) = n$. Thus we have

$$\beta = f(\alpha) = p(\alpha)q(\alpha) + r(\alpha) = r(\alpha).$$

It follows that β can be expressed as

$$\beta = b_0 + b_1\alpha + \cdots + b_{n-1}\alpha^{n-1}$$

with coefficients $b_i \in F$.

For linear independence, if

$$c_0 + c_1\alpha + \cdots + c_{n-1}\alpha^{n-1} = 0$$

for $c_i \in F$, then we have

$$g(x) = c_0 + c_1x + \cdots + c_{n-1}x^{n-1} \in F[x]$$

with $g(\alpha) = 0$. Noticing that $\deg(g(x)) \leq n - 1 < n = \deg(p(x))$ and $p(x)$ is the irreducible polynomial of α over F, we must have $g(x) = 0$. Therefore, $c_i = 0$, so the linear independence of the α^i is established. $\qquad\square$

We give an impressive example illustrating the theorem.

Example 4.2.5. *Is $p(x) = x^2 + x + 1 \in \mathbb{Z}_2[x]$ irreducible? Find a field that has a zero of $p(x)$.*

Solution. Since $p(0) = p(1) \neq 0$, $p(x)$ is irreducible over $\mathbb{Z}_2$. By Kronecker Theorem we know that the field $E = \mathbb{Z}_2[x]/\langle p(x)\rangle$ contains a zero α of $x^2 + x + 1$. It is clear that $|E| = 4$. Also we know that the extension field $\mathbb{Z}_2(\alpha)$ of $\mathbb{Z}_2$ contains the following four elements

$$0 + 0\alpha, 1 + 0\alpha, 0 + 1\alpha, 1 + 1\alpha,$$

that is, $0, 1, \alpha$ and $1 + \alpha$. This is a new finite field with four elements! Moreover, the addition and multiplication tables for this field are shown below. For example, to compute $(1 + \alpha)(1 + \alpha) \in \mathbb{Z}_2(\alpha)$, notice that $p(\alpha) = \alpha^2 + \alpha + 1 = 0$, we have

$$\alpha^2 = -\alpha - 1 = \alpha + 1.$$

Therefore,

$$(1 + \alpha)(1 + \alpha) = 1 + \alpha + \alpha + \alpha^2 = 1 + \alpha^2 = 1 + \alpha + 1 = \alpha.$$

+	0	1	α	$1 + \alpha$
0	0	1	α	$1 + \alpha$
1	1	0	$1 + \alpha$	α
α	α	$1 + \alpha$	0	1
$1 + \alpha$	$1 + \alpha$	α	1	0

$\cdot$	0	1	α	$1 + \alpha$
0	0	0	0	0
1	0	1	α	$1 + \alpha$
α	0	α	$1 + \alpha$	1
$1 + \alpha$	0	$1 + \alpha$	1	α

$\square$

Let F be a field and $f(x) \in F[x]$. We know by Kronecker Theorem that there exists an extension field E of F and an $\alpha \in E$ such that α is a root of $f(x)$. In order to discuss whether $f(x)$ has a multiple root, we need the following definition.

Definition 4.2.8. *Let F be a field and $f(x) = \sum_i a_i x^i \in F[x]$. Define the **derivative polynomial** of $f(x)$ to be $f'(x) = \sum_i i a_i x^{i-1}$.*

Note that one must be careful to realize that multiplication by i denotes multiplication by the image of i under the standard map from $\mathbb{Z}$ to F. In particular, it may be zero. For instance, the derivative polynomial of $x^p - 1 \in \mathbb{Z}_p[x]$ is zero.

The follow lemma is easy to prove or from calculus.

Lemma 4.2.9. *Let $f(x), g(x) \in F[x]$. Then we have*

$$[f(x)g(x)]' = f'(x)g(x) + f(x)g'(x).$$

Theorem 4.2.10. *Let $f(x) \in F[x]$. Then $f(x)$ has a multiple root in an extension field E if and only if $\gcd(f(x), f'(x)) \neq 1$.*

Proof. ($\Rightarrow$). Suppose that α is a multiple root of $f(x) \in F[x]$ in an extension field E. Then, there exists $g(x) \in E[x]$ such that $f(x) = (x - \alpha)^2 g(x)$. Using the product rule in Lemma 4.2.9, we see that $f'(x) = 2(x - \alpha)g(x) + (x - \alpha)^2 g'(x)$ and hence $1 \neq \gcd(f(x), f'(x))$.

($\Leftarrow$). Suppose that $\gcd(f(x), f'(x)) \neq 1$. Let $d(x) = \gcd(f(x), f'(x))$. Then there exists an extension field E in which $d(x)$ has a root, say α. Since $(x - \alpha)|d(x)$ and hence $f(x)$, there exists $h(x) \in E[x]$ such that $f(x) = (x - \alpha)h(x)$. So

$$f'(x) = h(x) + (x - \alpha)h'(x).$$

Since $(x - \alpha)$ divides $f'(x)$, it must also divide $h(x)$. But then $(x - \alpha)^2|f(x)$ in $E[x]$ as required. $\square$

4.3. Algebraic extensions and algebraic closure.

In this section we will prove that every field F has an extension E such that every nonconstant polynomial in $F[x]$ always has a root in E. Such a minimal extension field of F is the algebraic closure of F. We always assume that F is a field in this section.

Definition 4.3.1. *Let $F \leq E$ be fields. Then E is called is an* **algebraic extension** *of F if every element in E is algebraic over F.*

We first have the following theorem which says that any finite extension field is an algebraic extension.

Theorem 4.3.2. *Let $F \leq E$ be fields with $[E : F] < \infty$. Then E is an algebraic extension of F.*

Proof. Assume that $[E : F] = n$. For any $\alpha \in E$, then $\{1, \alpha, \cdots, \alpha^n\}$ is linearly dependent over F, so there exist $a_i \in F$, $0 \leq i \leq n$, such that

$$a_n\alpha^n + \cdots + a_1\alpha + a_0 = 0,$$

and not all $a_i = 0$. Then $f(x) = a_nx^n + \cdots + a_1x + a_0 \in F[x]$ is nonzero, and $f(\alpha) = 0$. Thus α is an algebraic element over F. $\square$

Corollary 4.3.3. *Let $F \leq E$ be fields and $\alpha \in E$ be algebraic over F, and $\beta \in F(\alpha)$. Then $\deg(\beta, F)|\deg(\alpha, F)$.*

Proof. Note that $\deg(\alpha, F) = [F(\alpha) : F]$ and $\deg(\beta, F) = [F(\beta) : F]$. Since $F \leq F(\beta) \leq F(\alpha)$, from Theorem 4.1.4 we see that $[F(\beta) : F]|[F(\alpha) : F]$. $\square$

Example 4.3.1. *Find a basis for $\mathbb{Q}(2^{1/2}, 2^{1/3})$ over $\mathbb{Q}$, and show that $\mathbb{Q}(2^{1/2}, 2^{1/3}) = \mathbb{Q}(2^{1/6})$.*

Solution. Since $\deg(2^{1/2}, \mathbb{Q}) = 2$ and $2 \nmid 3 = \deg(2^{1/3}, \mathbb{Q})$, we see that $2^{1/2} \notin \mathbb{Q}(2^{1/3})$. Then $x^2 - 2$ is irreducible over $\mathbb{Q}(2^{1/3})$ and

$$[\mathbb{Q}(2^{1/3}, 2^{1/2}) : \mathbb{Q}(2^{1/3})] = 2.$$

So $\{1, 2^{1/3}, 2^{2/3}\}$ is a basis for $\mathbb{Q}(2^{1/3})$ over $\mathbb{Q}$ and $\{1, 2^{1/2}\}$ is a basis for $\mathbb{Q}(2^{1/3}, 2^{1/2})$ over $\mathbb{Q}(2^{1/3})$. Moreover, by Theorem 4.2.7, $\{1, 2^{1/2}, 2^{1/3}, 2^{5/6}, 2^{2/3}, 2^{7/6}\}$ is a basis for $\mathbb{Q}(2^{1/2}, 2^{1/3})$ over $\mathbb{Q}$.

Since $2^{7/6} = 2(2^{1/6})$, we see that $2^{1/6} \in \mathbb{Q}(2^{1/2}, 2^{1/3})$. Note that $2^{1/6}$ is a zero of $x^6 - 2$, which is irreducible over $\mathbb{Q}$ by Schönemann-Eisenstein Criterion. Since

$$\mathbb{Q} \leq \mathbb{Q}(2^{1/6}) \leq \mathbb{Q}(2^{1/2}, 2^{1/3}),$$

we have

$$6 = [\mathbb{Q}(2^{1/2}, 2^{1/3}) : \mathbb{Q}] = [\mathbb{Q}(2^{1/2}, 2^{1/3}) : \mathbb{Q}(2^{1/6})][Q(2^{1/6}) : \mathbb{Q}]$$

$$= 6[\mathbb{Q}(2^{1/2}, 2^{1/3}) : \mathbb{Q}(2^{1/6})].$$

Thus, $[\mathbb{Q}(2^{1/2}, 2^{1/3}) : \mathbb{Q}(2^{1/6})] = 1$, and $\mathbb{Q}(2^{1/2}, 2^{1/3}) = \mathbb{Q}(2^{1/6})$. $\square$

The previous example shows that it is possible for an extension $F(\alpha_1, \cdots, \alpha_n)$ for $n > 1$ of a field F to be a simple extension.

Theorem 4.3.4. *Let E be an algebraic extension of a field F. Then there exist $\alpha_1, \cdots, \alpha_n \in E$ such that $E = F(\alpha_1, \cdots, \alpha_n)$ if and only if E is a finite extension of F.*

Proof. ($\Rightarrow$). Suppose that $E = F(\alpha_1, \cdots, \alpha_n)$ for some elements $\alpha_i \in E$. Since E is an algebraic extension of F, each α_i is algebraic over F. So α_1 is algebraic over F, and moreover, α_j is algebraic over $F(\alpha_1, \cdots, \alpha_{j-1})$ for $j = 2, \cdots, n$. So $[F(\alpha_1, \cdots, \alpha_j) : F(\alpha_1, \cdots, \alpha_{j-1})] < \infty$. For the sequence of finite extensions

$$F \leq F(\alpha_1) \leq F(\alpha_1, \alpha_2), \cdots \leq F(\alpha_1, \cdots, \alpha_n) = E,$$

we know that E is a finite extension of F (Corollary 4.1.5).

($\Leftarrow$). Suppose that E is a finite algebraic extension of F, i.e., $[E : F] = n < \infty$. Take a basis $\{\alpha_1, \cdots, \alpha_n\}$ of E over F. Clearly

$$F(\alpha_1, \cdots, \alpha_n) = E.$$

$\square$

We have observed that if E is an extension of a field F and $\alpha, \beta \in E$ are algebraic over F, then so are $\alpha + \beta, \alpha\beta, \alpha - \beta$, and α/β if $\beta \neq 0$. This follows also from the following theorem.

Theorem 4.3.5. *Let $F \leq E$ be fields. Then*

$$\overline{F}_E = \{\alpha \in E | \alpha \text{ is algebraic over } F\} \leq E,$$

*called the **algebraic closure** of F in E.*

Proof. Let $\alpha, \beta \in \overline{F}_E$. Then $F(\alpha, \beta)$ is a finite extension of F, and every element of $F(\alpha, \beta)$ is algebraic over F, that is, $F(\alpha, \beta) \subset \overline{F}_E$. Thus $\overline{F}_E$ contains $\alpha + \beta, \alpha\beta, \alpha - \beta$, and also contains α/β for $\beta \neq 0$, so $\overline{F}_E \leq E$. $\quad\square$

Corollary 4.3.6. *The set of all algebraic numbers forms a field.*

Proof. This corollary follows from the previous Theorem by taking $F = \mathbb{Q}$ and $E = \mathbb{C}$. $\quad\square$

It is well known that the complex numbers have the property that every nonconstant polynomial in $\mathbb{C}[x]$ has a zero in $\mathbb{C}$. This is known as the Fundamental Theorem of Algebra. We will give a proof for this theorem in Theorem 6.2.1. We now give a name for such fields in general.

Definition 4.3.7. *Let F be a field. F is called **algebraically closed** if every nonconstant polynomial in $F[x]$ has a zero in F.*

Theorem 4.3.8. *A field F is algebraically closed if and only if every nonconstant polynomial in $F[x]$ is a product of degree one polynomials in $F[x]$.*

Proof. ($\Rightarrow$). Let F be algebraically closed and $f(x) \in F[x]$ with $\deg(f(x)) = n \geq 1$. We prove this by induction on n. If $n = 1$ this is trivial. If $n > 1$, then $f(x)$ has a zero $\alpha \in F$. So $x - \alpha$ is a factor of $f(x)$, and $f(x) = (x - a)g(x)$ for some $g(x) \in F[x]$. Note that $\deg(g(x)) = n - 1$. By inductive hypothesis $g(x)$ is a product of degree one polynomials in $F[x]$. Hence $f(x)$ is a product of degree one polynomials in $F[x]$.

($\Leftarrow$). Suppose that every nonconstant polynomial $f(x)$ of $F[x]$ can be written as a product of linear factors, say

$$f(x) = (a_1 x - b_1)(a_2 x - b_2) \cdots (a_n x - b_n).$$

Then we have b_i/a_i, $1 \leq i \leq n$, are all zeros of $f(x)$. Thus F is algebraically closed. $\quad\square$

Corollary 4.3.9. *Let F be an algebraically closed field. Then there is no algebraic extension E of F such that $F < E$.*

Proof. Let E be an algebraic extension of F and $\alpha \in E$. Since F is algebraically closed, we have $\mathrm{irr}(\alpha, F) = x - \alpha$. Thus $\alpha \in F$, it follows that $F = E$. $\quad\square$

If a field F has an algebraic extension $\overline{F}$ which is algebraically closed, then $\overline{F}$ will certainly be a maximal algebraic extension of F, since $\overline{F}$ is algebraically closed, it can have no proper algebraic extensions. Such an extension of F is called an **algebraic closure of F**.

The proof of the following lemma is left as Exercise (4).

Lemma 4.3.10. *Let $K \leq F \leq E$ be fields. If E is algebraic over F and F is algebraic over K, then E is algebraic over K.*

Theorem 4.3.11. *Every field F has an algebraic closure $\overline{F}$.*

Proof. Let
$$P = \{p_j(x) \in F[x] : j \in J,$$

each $p_j(x)$ is monic and irreducible over $F\}$,

that is, P is the set of all irreducible monic polynomials in $F[x]$. Construct a polynomial ring $R = F[x_j : j \in J]$ in infinitely many variables x_j. Now let $I = \langle p_j(x_j) : j \in J \rangle \trianglelefteq R$.

First $I \neq R$. Otherwise if $I = R$ then $1 \in I$, so one can write

$$1 = \sum_{j=1}^{n} a_j p_j(x_j) \tag{4.1}$$

for some $a_i \in R$. Using Theorem 4.2.1 we can have an extension field E of F containing all the roots for the polynomial $p_1 p_2 \cdots p_n$, and choose a root $\alpha_j \in E$ for each p_j, $1 \leq i \leq n$. We now consider the evaluation map
$$\phi : R \to E,$$

$$x_j \to \begin{cases} \alpha_j, & \text{if } 1 \leq j \leq n \\ 0, & \text{otherwise.} \end{cases}$$

The Equation (4.1) becomes $1 = 0$, which is nonsense. This proves that $I \neq R$.

Consider $S = \{K \trianglelefteq R : I \subset K\}$ as a partially ordered set under the set inclusion. Let $\mathcal{C}$ be a chain in S. One can easily show that $\cup_{K \in \mathcal{C}} K$ is an upper bound of $\mathcal{C}$ in S. By Kuratowski-Zorn Lemma there's a maximal ideal N of R with $I \subset N \subset R$. Define $F_1 = R/N$ and we have the canonical map $\phi_F : F \to F_1$. Furthermore every polynomial $f(x) \in F[x]$ of degree 1 or more has a root in F_1! We identify F with $\phi_F(F)$, that is, F is a subfield of F_1. Since $p_j(x_j + N) = 0 \in F_1 = R/N$, we know that F_1 is an algebraic extension of F (Lemma 4.3.10).

It would be great if we were now done. Unfortunately we are not yet there. The problem is that F_1 has the property that every polynomial in $F[x]$ of positive degree has a root. However there may be polynomials of positive degree in the larger ring $F_1[x]$ that do not have roots in F_1.

In this manner, if we start with F_1 then we build a field F_2 containing F_1 such that every element of $F_1[x]$ of positive degree has a root in F_2. Similarly, F_2 is an algebraic extension of F_1 and hence of F (Lemma 4.3.10). And so on. We continue, getting an infinite collection of algebraic extensions

$$F \subset F_1 \subset \cdots \subset F_k \subset \cdots.$$

Now, we let $\overline{F} = \cup_{k \in \mathbb{N}} F_k$. Then $\overline{F}$ is an algebraic extension field of F (Lemma 4.3.10). Because any $f \in \overline{F}[x]$ of degree 1 or more will have each coefficient in some F_k for k large enough, so $f(x)$ has a root in F_{k+1} and hence in $\overline{F}$. Then $\overline{F}$ will be an algebraically closed field, and is an algebraic extension of F. Thus $\overline{F}$ is an algebraic closure of F. □

Later in Corollary 5.2.3 we will show that each field F has a unique algebraic closure $\overline{F}$ up to isomorphisms.

Definition 4.3.12. *Let $F \leq E$ be fields. A subset S of a field E is* **algebraically independent** *over F if the elements of S do not satisfy any non-trivial polynomial equation with coefficients in F. A maximal algebraically independent subset of E over F is called a* **transcendence basis** *of E over F.*

Using Kuratowski-Zorn Lemma one can show that transcendence bases of E over F always exist and all have the same cardinality which is called the **transcendence degree** of E over F. It is interesting to know the following theorem that is named for Ferdinand von Lindemann (1852–1939) and Karl Weierstrass (1815–1897). Lindemann proved in 1882 that e^a is transcendental for every non-zero algebraic number a, thereby establishing that π is transcendental. Weierstrass proved the above more general statement in 1885. Here we do not provide a proof which can be found in [J].

Theorem 4.3.13 (Lindemann-Weierstrass Theorem). *If $\alpha_1, \alpha_2, \cdots, \alpha_n$ are algebraic numbers that are linearly independent over $\mathbb{Q}$, then $e^{\alpha_1}, e^{\alpha_2}, \cdots, e^{\alpha_n}$ are algebraically independent over $\mathbb{Q}$.*

Open Problem. Although both π and e are known to be transcendental, it is not known whether π and e are algebraically independent over $\mathbb{Q}$. Even it is not known whether $e - \pi$ or $e + \pi$ is irrational.

Example 4.3.2. *Prove that $x^2 - 3$ is irreducible over $\mathbb{Q}(\sqrt[3]{2})$.*

Solution. If $x^2 - 3$ were reducible over $\mathbb{Q}(\sqrt[3]{2})$, then it would factor into linear factors over $\mathbb{Q}(\sqrt[3]{2})$, so $\sqrt{3}$ would lie in the field $\mathbb{Q}(\sqrt[3]{2})$, and we would have $\mathbb{Q}(\sqrt{3}) \leq \mathbb{Q}(\sqrt[3]{2})$. But then

$$[\mathbb{Q}(\sqrt[3]{2}) : \mathbb{Q}] = [\mathbb{Q}(\sqrt[3]{2}) : \mathbb{Q}(\sqrt{3})][\mathbb{Q}(\sqrt{3}) : \mathbb{Q}].$$

This equation is impossible because $[\mathbb{Q}(\sqrt[3]{2}) : \mathbb{Q}] = 3$ while $[\mathbb{Q}(\sqrt{3}) : \mathbb{Q}] = 2$. This is impossible. So $x^2 - 3$ is irreducible over $\mathbb{Q}(\sqrt[3]{2})$. $\square$

Example 4.3.3. *Let $a, b \in \mathbb{Q}$. If $\sqrt{a} + \sqrt{b} \neq 0$, show that $\mathbb{Q}(\sqrt{a} + \sqrt{b}) = \mathbb{Q}(\sqrt{a}, \sqrt{b})$.*

Solution. If $a = b$ the result is clear; we assume $a \neq b$. It is obvious that $\mathbb{Q}(\sqrt{a} + \sqrt{b}) \subset \mathbb{Q}(\sqrt{a}, \sqrt{b})$.
We now show that $\mathbb{Q}(\sqrt{a}, \sqrt{b}) \subset \mathbb{Q}(\sqrt{a} + \sqrt{b})$. Let $\alpha = \frac{a-b}{\sqrt{a}+\sqrt{b}} \in \mathbb{Q}(\sqrt{a} + \sqrt{b})$. Now $\alpha = \sqrt{a} - \sqrt{b}$. Thus $\mathbb{Q}(\sqrt{a} + \sqrt{b})$ contains $\frac{1}{2}[\alpha + (\sqrt{a} + \sqrt{b})] = \sqrt{a}$ and hence also contains $(\sqrt{a} + \sqrt{b}) - \sqrt{a} = \sqrt{b}$. Thus $\mathbb{Q}(\sqrt{a}, \sqrt{b}) \subset \mathbb{Q}(\sqrt{a} + \sqrt{b})$. $\square$

Example 4.3.4. *Find a basis for $\mathbb{Q}(\sqrt{2}, \sqrt{3})$ over $\mathbb{Q}$.*

Solution. Since $\sqrt{3} \notin \mathbb{Q}(\sqrt{2})$, we see that

$$[\mathbb{Q}(\sqrt{2}, \sqrt{3}) : \mathbb{Q}] = [\mathbb{Q}(\sqrt{2}, \sqrt{3}) : \mathbb{Q}(\sqrt{3})][\mathbb{Q}(\sqrt{3}) : \mathbb{Q}] = 2 \times 2 = 4.$$

Since $\mathbb{Q}(\sqrt{2} + \sqrt{3}) = \mathbb{Q}(\sqrt{2}, \sqrt{3})$ and $\sqrt{2} + \sqrt{3}$ is a zero of $x^4 - 10x^2 + 1$, then $\mathrm{irr}(\sqrt{2} + \sqrt{3}, \mathbb{Q}) = x^4 - 10x^2 + 1$. (This is a method to show that this polynomial is irreducible.) Consequently, $\{1, \sqrt{3}\}$ is a basis for $\mathbb{Q}(\sqrt{2}, \sqrt{3}) = (\mathbb{Q}(\sqrt{2}))(\sqrt{3})$ over $\mathbb{Q}(\sqrt{2})$. This shows that $\{1, \sqrt{2}, \sqrt{3}, \sqrt{6}\}$ is a basis for $\mathbb{Q}(\sqrt{2}, \sqrt{3})$ over $\mathbb{Q}$. $\square$

4.4. Finite fields.

We shall now apply the established results in the extension field theory to determine the structure of all finite fields. Observe that if F is a finite field, then $\mathrm{char}(F) = p$ is a prime and the prime field of F can be identified with $\mathbb{Z}_p$. In the usual way we may regard F as a vector space over $\mathbb{Z}_p$. Assume that $[F : \mathbb{Z}_p] = n$, then we have a basis $\{\alpha_1, \cdots, \alpha_n\}$ of F over $\mathbb{Z}_p$, every element on F can be written in one and only one way as a linear combination

$$a_1\alpha_1 + a_2\alpha_2 + \cdots + a_n\alpha_n,$$

where $a_i \in \mathbb{Z}_p$. Since each a_i may be any of the p elements of $\mathbb{Z}_p$, the total number of such distinct linear combinations of the α_i is p^n, that is, $|F| = p^n$. The same method shows that if $F \le E$, $[E : F] = n$ and $|F| = q$ then $|E| = q^n$. Thus we have the basic facts on finite fields.

Theorem 4.4.1. *Let F be a finite field with* $\mathrm{char}(F) = p$.

(1). $|F| = p^n$ for some positive integer n.

(2). If $F \le E$, $[E : F] = n$, and $|F| = q$, then $|E| = q^n$.

We shall now consider the structure of the multiplicative group of nonzero elements of a finite field.

Theorem 4.4.2. *Let E be a finite field. Then the multiplicative group $(E^*, \cdot)$ of E is cyclic.*

Proof. This follows from Corollary 1.6.5. □

Corollary 4.4.3. *Any finite extension E of a finite field F is a simple extension of F.*

Proof. If α is a generator for the cyclic group E^* of nonzero elements of E, we see that $E = F(\alpha)$. □

Lemma 4.4.4. *Let F be a field with* $\mathrm{char}(F) = p$. *Then for all $\alpha, \beta \in F$ and all $n \in \mathbb{Z}^+$ we have*

$$(\alpha + \beta)^{p^n} = \alpha^{p^n} + \beta^{p^n}.$$

Proof. Applying the binomial theorem to $(\alpha + \beta)^p$, we have

$$(\alpha + \beta)^p = \alpha^p + (p \cdot 1)\alpha^{p-1}\beta + \frac{p(p-1)}{2}$$
$$\cdot 1\alpha^{p-2}\beta^2 + \cdots + (p \cdot 1)\alpha\beta^{p-1} + \beta^p$$
$$= \alpha^p + 0\alpha^{p-1}\beta + 0\alpha^{p-2}\beta^2 + \cdots + 0\alpha\beta^{p-1} + \beta^p$$
$$= \alpha^p + \beta^p.$$

By induction on n, suppose that we have $(\alpha+\beta)^{p^{n-1}} = \alpha^{p^{n-1}} + \beta^{p^{n-1}}$. Then

$$(\alpha + \beta)^{p^n} = [(\alpha + \beta)^{p^{n-1}}]^p = (\alpha^{p^{n-1}} + \beta^{p^{n-1}})^p = \alpha^{p^n} + \beta^{p^n}.$$

□

Let $\overline{\mathbb{Z}}_p$ be an algebraic closure of the field $\mathbb{Z}_p$, $n \in \mathbb{N}$, and
$$F_{p^n} = \{a \in \overline{\mathbb{Z}}_p : a^{p^n} = a\}.$$
Theorem 4.4.5. *Let p be a prime and $n \in \mathbb{N}$. Then F_{p^n} is a field of order p^n.*

Proof. Using the previous lemma we can easily show that F_{p^n} is a subfield of $\overline{\mathbb{Z}}_p$ of order at most p^n since elements in F_{p^n} are solutions of the polynomial $f(x) = x^{p^n} - x$. From Theorem 4.2.10 and the fact that $\gcd(f(x), f'(x)) = 1$ we see that the number of solutions of $x^{p^n} - x$ is p^n. □

Thus, for each prime p we have the sequence of finite fields
$$\mathbb{Z}_p = F_p < F_{p^2} < F_{p^3} < \cdots < F_{p^r} < \cdots < \overline{\mathbb{Z}}_p.$$
This tells us that $\overline{F}_p = \cup_{k=1}^{\infty} F_{p^k}$. See Exercise (20).

Definition 4.4.6. *Let E be a field and $\alpha \in E$. Then α is called an n-th root of unity if $\alpha^n = 1$. It is called a primitive n-th root of unity if $\alpha^n = 1$ and $\alpha^m \neq 1$ for $0 < m < n$.*

We know that the nonzero elements of a finite field of p^n elements are all $(p^n - 1)$-th roots of unity.

Example 4.4.1. *Find the generators for $(\mathbb{Z}_{11}^*, \cdot)$, i.e., all primitive 10th roots of unity in the field $(\mathbb{Z}_{11}, +, \cdot)$, and all primitive 5th roots of unity.*

Solution. Consider the finite field $\mathbb{Z}_{11}$. We know that $(\mathbb{Z}_{11}^*, \cdot)$ is cyclic of order 10. Let us find all generators of $\mathbb{Z}_{11}^*$. Noticing that $|\mathbb{Z}_{11}^*| = 10$, we see that $\text{ord}(2)|10$. It follows that $\text{ord}(2) = 2, 5$, or 10. Since
$$2^2 = 4, 2^4 = 4^2 = 5, \text{ and } 2^5 = (2)(5) = 10 = -1,$$
so $\text{ord}(2) = 10$, and $\mathbb{Z}_{11}^* = \langle 2 \rangle$, that is, 2 is a primitive 10th root of unity in $\mathbb{Z}_{11}$.

By the theory of cyclic groups, we know that $\text{ord}(2^n) = \frac{10}{\gcd(n,10)}$. Then all the generators of Z_{11}^*, are of the form 2^n, where $\gcd(n, 10) = 1$. Thus, these elements are
$$2^1 = 2, 2^3 = 8, 2^7 = 7, 2^9 = 6.$$

The primitive 5th roots of unity in $\mathbb{Z}_{11}$ are of the form 2^m, where $\gcd(m, 10) = 2$, that is,
$$2^2 = 4, 2^4 = 5, 2^6 = 9, 2^8 = 3.$$
The primitive square root of unity in $\mathbb{Z}_{11}$ is $2^5 = 10 = -1$. □

Corollary 4.4.7. *Let F be a finite field. Then for every positive integer n, there is an irreducible polynomial in $F[x]$ of degree n.*

Proof. Let $\mathrm{char}(F) = p$, $|F| = p^r$ and $\overline{F}$ be the algebraic closure of F. Then there is a field $K \leq \overline{F}$ consisting precisely of the p^{rn} zeros of $x^{p^{rn}} - x$. Notice that all elements in F are zeros of $x^{p^r} - x$, by Theorem 4.4.5 we know that $F \leq K$ and $[K : F] = n$. Since K is simple over F (Corollary 4.4.3), so $K = F(\beta)$ for some $\beta \in K$. Hence, the irreducible polynomial $\mathrm{irr}(\beta, F)$ is of degree $n = [K : F]$ (Theorem 4.2.7). $\qquad\square$

Theorem 4.4.8. *Let p be a prime and let $n \in \mathbb{N}$. If E and E' are fields of order p^n, then $E \cong E'$.*

Proof. We may regard both E and E' as extensions of the prime field $\mathbb{Z}_p$ up to isomorphism. Then E is a simple extension of $\mathbb{Z}_p$ of degree n, say $E = \mathbb{Z}_p(\alpha)$. Let $f(x) = \mathrm{irr}(\alpha, \mathbb{Z}_p)$ which has degree n. By considering the evaluation homomorphism $\phi_\alpha : \mathbb{Z}_p[x] \to E$ we see that $E \cong \mathbb{Z}_p[x]/\langle f(x) \rangle$. Because elements of E are zeros of $x^{p^n} - x$, we see that $f(x)|x^{p^n} - x$ in $\mathbb{Z}_p[x]$. Because E' also consists of all zeros of $x^{p^n} - x$, then E' also contains zeros of irreducible $f(x) \in \mathbb{Z}_p[x]$, say $\alpha' \in E'$ with $f(\alpha') = 0$. Thus $E' = \mathbb{Z}_p(\alpha')$ since they have the same size. By considering the evaluation homomorphism $\phi_{\alpha'} : \mathbb{Z}_p[x] \to E'$ we see that $E' \cong \mathbb{Z}_p[x]/\langle f(x) \rangle$ since $\ker(\phi_{\alpha'}) = \langle f(x) \rangle$. Therefore $E' \cong E$. $\qquad\square$

As a classical application, we will use Gaussian integers and properties of finite fields to prove the following result in Number Theory.

Theorem 4.4.9. *Let p be an odd prime in $\mathbb{Z}$. Then $p = a^2 + b^2$ for some $a, b \in \mathbb{Z}$ if and only if $p \equiv 1 \pmod{4}$.*

Proof. ($\Rightarrow$). Suppose that $p = a^2 + b^2$. Since p is odd, then a and b cannot be both even or both odd. We may assume that $a = 2r$ and $b = 2s + 1$, then $a^2 + b^2 = 4r^2 + 4(s^2 + s) + 1$, so $p \equiv 1 \pmod{4}$.

($\Leftarrow$). Assume that $p \equiv 1 \pmod{4}$. From Theorem 4.4.2 we know that the multiplicative group of nonzero elements of the finite field $\mathbb{Z}_p$ is cyclic of order $p - 1$. Since $4|p - 1$, we see that the group $(\mathbb{Z}_p^*, \cdot)$ contains an element n of order 4. Then n^2 has multiplicative order 2. So $n^2 = -1$ in $\mathbb{Z}_p$. Thus $n^2 \equiv -1 \pmod{p}$, so $p|n^2 + 1$ in $\mathbb{Z}$.

Next we work within in the Euclidean domain $\mathbb{Z}[i]$ (Theorem 2.5.4). We see that $p|n^2 + 1 = (n + i)(n - i)$ in $\mathbb{Z}[i]$. If p is irreducible in $\mathbb{Z}[i]$, then $p|n + i$ or $p|n - i$, which is impossible. Thus p is not irreducible in $\mathbb{Z}[i]$.

Let $p = (a + bi)(c + di)$ where neither $a + bi$ nor $c + di$ is a unit. Then $a^2 + b^2 \neq 1$ nor $c^2 + d^2 \neq 1$. Taking norms, we have $p^2 = (a^2 + b^2)(c^2 + d^2)$. So $p = a^2 + b^2 = c^2 + d^2$, which completes our proof. □

To conclude this chapter we prove the following beautiful result in Number Theory using Gaussian integers $\mathbb{Z}[i]$.

Theorem 4.4.10. *Let $n = p_1^{n_1} p_2^{n_2} \cdots p_k^{n_k} \in 1 + \mathbb{N}$ where $p_1, p_2, \cdots, p_k$ are pairwise distinct primes with each $n_j \in \mathbb{N}$. Then $n = a^2 + b^2$ for some $a, b \in \mathbb{Z}$, if and only if, $p_j \equiv 3 \pmod 4$ implies n_j is even.*

Proof. ($\Rightarrow$). We have $n = a^2 + b^2 = (a + ib)(a - ib) = N(a + ib)$. Let $z = a + ib \in \mathbb{Z}[i]$. Write $z = \alpha_1 \cdots \alpha_q$ as a product of irreducibles in $\mathbb{Z}[i]$. By Exercise (34), we see that

   (a). $\alpha_j = \pm 1 \pm i$,
   (b). $\alpha_j = p \equiv 3 \pmod 4$,
   (c). $N(\alpha_j) = p \equiv 1 \pmod 4$,

where p is a prime number in $\mathbb{Z}$. We now take the norm for $z = a + ib$ to obtain

$$n = a^2 + b^2 = N(z) = N(\alpha_1)N(\alpha_2) \cdots N(\alpha_q).$$

If $\alpha_j = \pm 1 \pm i$, we know that $N(\alpha_j) = 2$; if $\alpha_j = p \equiv 3 \pmod 4$, we know that $N(\alpha_j) = p^2$; the remaining case for α_j is $N(\alpha_j) = p \equiv 1 \pmod 4$. So $p_j \equiv 3 \pmod 4$ implies n_j is even.

($\Leftarrow$). Let $n = p_1^{n_1} p_2^{n_2} \cdots p_k^{n_k}$ be a product of distinct primes such that $p_j \equiv 3 \pmod 4$ implies n_j is even. By Theorem 4.4.9, we know that each $p_j^{n_j}$ is a sum of integer squares. From Exercise (32), we see that $n = a^2 + b^2$ for some $a, b \in \mathbb{Z}$. □

Example 4.4.2. *Can you write the integer 330000 as a sum of two integer squares?*

Solution. Note that $330000 = 3 \cdot 11 \cdot 100^2$. Since the power 3 is odd, by Theorem 4.4.10 we cannot write the integer 330000 as a sum of two integer squares. □

For $n \in \mathbb{N}$ with $n = a^2 + b^2$ for some $a, b \in \mathbb{Z}$, note that n may have more than one expressions as a sum of two integer squares. For example,

$$65 = 8^2 + 1^2 = 4^2 + 7^2.$$

4.5. Exercises.

(1) Calculate the irreducible polynomial of $\sqrt[3]{2}$ and $1 + \sqrt[3]{2} + \sqrt[3]{4}$ over $\mathbb{Q}$.

(2) Show that $\mathbb{Q}[x]/\langle x^2 - 1 \rangle$ is not an integral domain, but $\mathbb{Q}[x]/\langle x^2 + 1 \rangle$ is.

(3) Use Kuratowski-Zorn Lemma to show that every proper ideal of a ring R with unity is contained in some maximal ideal.

(4) Prove Lemma 4.3.10.

(5) Let E be a finite extension field of a field F. Let D be an integral domain such that $F \subset D \subset E$. Show that D is a field.

(6) Find the degree and a basis for the given field extensions: $\mathbb{Q}(\sqrt[3]{2}, \sqrt[3]{6}, \sqrt[3]{24})$ over $\mathbb{Q}, \mathbb{Q}(\sqrt[3]{2}, \sqrt{3})$ over $\mathbb{Q}$.

(7) Let E be an extension field of a field F and $[E : F]$ be a prime. For any $\alpha \in E \setminus F$, show that $E = F(\alpha)$.

(8) Let E be an extension field of F and $\alpha \in E$ be algebraic of odd degree over F. Show that α^2 is algebraic of odd degree over F and $F(\alpha) = F(\alpha^2)$.

(9) Let $f(x)$ be an irreducible polynomial in $\mathbb{Z}_p[x]$. Show that $f(x)$ is a divisor of $x^{p^n} - x$ for some n.

(10) Let $c \in F$, where F is a field of characteristic $p > 0$. Prove that $x^p - x - c$ is irreducible in $F[x]$ if and only if $x^p - x - c$ has no root in F. Show this is false if F is of characteristic 0.

(11) Let $f(x) = x^p - x - c$ where p is a prime not dividing $c \in \mathbb{Z}$. Show that $f(x)$ is irreducible over $\mathbb{Q}$. (Hint: try the following steps: Show that if $f(x)$ is irreducible over $\mathbb{Z}_p$, then it is irreducible over $\mathbb{Q}$. Show that $f(x)$ does not have a root in $\mathbb{Z}_p$. Then consider the previous exercise.)

(12) Under what conditions on q is the polynomial $x^2 + x + 1$ irreducible over a finite field F with q elements? (Hint: consider the multiplicative group of nonzero elements of F.)

(13) Find the conditions on $a \in \mathbb{C}$ such that $x^5 - 5x + a = 0$ has multiple roots.

(14) Let F be an algebraically closed field. Find conditions on $a \in F$ such that the equation $x^5 + 5ax + 4a = 0$ has no multiple roots in F.

(15) Let p be a prime. Show that a finite field of p^n elements has exactly one subfield of p^m elements for each divisor m of n.

(16) Find a primitive root of unity of order 6 in F_7.

(17) Find the number of primitive root of unity of order 48 in F_{7^2}.

(18) Find the number of primitive roots of unity of order 24 in F_{7^2}.

(19) Let p be a prime, $m \in \mathbb{N}$ such that $p \nmid m$. Show that there is $n \in \mathbb{N}$ such that $m | p^n - 1$.

(20) Let p be a prime, $F_p < F_{p^2} < F_{p^3} < \cdots < F_{p^r} < \cdots$. Show that $\overline{F}_p = \cup_{k=1}^{\infty} F_{p^k}$.

(21) Let F be a field consisting of 13^{11} elements. Find the number of subfields of F.

(22) Let F be a field consisting of 13^{11} elements and $\alpha \in F$. Find $\deg(\alpha, \mathbb{Z}_{13})$ where $\mathbb{Z}_{13}$ is the prime subfield of F.

(23) Let F be a field consisting of 13^{11} elements. How many distinct irreducible polynomials of degree 11 over $\mathbb{Z}_{13}$.

(24) Let p be a prime, F be the extension field of $\mathbb{Z}_p$ of degree n, consisting of all zeros of $x^{p^n} - x$. For each $d|n$ let $f_{d,1}(x)$, $f_{d,2}(x), \cdots, f_{d,r_d}(x)$ be all the distinct irreducible monic polynomials of degree d in $\mathbb{Z}[x]$. Show that

$$x^{p^n} - x = \prod_{d|n} \prod_{j=1}^{r_d} f_{d,j}(x).$$

(25) Let F be the field $\mathbb{Z}_2[x]/\langle x^6 + x + 1 \rangle$ of order $2^6 = 64$.
   (a). Calculate $x^7, x^9, x^{14}(= (x^7)^2)$ and $x^{21}(= (x^7)^3)$ in F as $\mathbb{Z}_2$-linear combinations of $1, x, \cdots, x^5$.
   (b). Find the order of x in the group $\mathcal{U}(F)$ of units of F.

(26) There are two irreducible cubics $x^3 + x + 1$ and $x^3 + x^2 + 1$ in $\mathbb{Z}_2[x]$. Show that $\mathbb{Z}_2[x]/\langle x^3 + x + 1 \rangle \cong \mathbb{Z}_2[x]/\langle x^3 + x^2 + 1 \rangle$.

(27) Let p be a prime, and let $f(x), g(x) \in \mathbb{Z}_p[x]$ be irreducible polynomials of order 6. Show that $\mathbb{Z}_p[x]/\langle f(x) \rangle \cong \mathbb{Z}_p[x]/\langle g(x) \rangle$.

(28) Find the number of different subfields of F_{7^8}.

(29) Let p be a prime. Let E be a field extension of F of degree p. If $a \in E \setminus F$, prove that the irreducible polynomial of a over F has degree p.

(30) Prove that $e^2 - 2$ is algebraic over $\mathbb{Q}(e^3)$. Find $\mathrm{irr}(e^2 - 2, \mathbb{Q}(e^3))$.

(31) Let F be a finite field with $\text{ch}(F) \neq 2$. For any $k \in \mathbb{Z}^+$ show that
$$\sum_{r \in F} r^k = 0,$$
where $0^0 = 1$. (Hint: Use Theorem 4.4.2.)

(32) Can you easily show that $(a^2 + b^2)(c^2 + d^2) = (ac - bd)^2 + (ad + bc)^2$ for any $a, b, c, d \in \mathbb{Z}$ without expanding both sides?

(33) If $p \in \mathbb{N}$ is prime with $p \equiv 3 \pmod{4}$, show that p is irreducible in the Gaussian integers $\mathbb{Z}[i]$.

(34) Let $a + bi \in \mathbb{Z}[i]$ where $a, b \in \mathbb{Z}$. Show that $a + bi$ is irreducible in $\mathbb{Z}[i]$ if and only if one of the following holds:
(a). $a + bi = \pm 1 \pm i$,
(b). $b = 0$ and $a \equiv 3 \pmod{4}$ is a prime, or
(c). $a^2 + b^2 \equiv 3 \pmod{4}$ is a prime.

(35) Can you write the integer 690000 as a sum of two integer squares?

(36) Can you write the integer 1146600 as a sum of two integer squares?

(37) Show that $\mathbb{Z}[x^2] \cap \mathbb{Z}[x^2 - x] = \mathbb{Z}$ in $\mathbb{Z}[x]$.

(38) Using Theorem 4.3.13 show that π is a transcendental number.

(39) If α is a non-zero algebraic number, using Theorem 4.3.13 show that $\sin(\alpha), \cos(\alpha), \tan(\alpha)$ are transcendental numbers. (Note that $\sin^2(\alpha) + \cos^2(\alpha) = 1$.)

(40) If $\alpha \neq 1$ is a positive algebraic number, show that $\ln(\alpha)$ is a transcendental number.

5. Automorphisms of Fields

This chapter is a preparation for Galois Theorem. We will study automorphism groups for various different extension fields. In this chapter we always assume that E, F are fields, $\overline{F}$ is the algebraic closure of F.

5.1. Automorphisms.

Definition 5.1.1. Let $F \leq E$ be fields, and $\alpha, \beta \in E$ be algebraic over F. We say that $\alpha, \beta \in E$ to be **conjugate over** F if $\mathrm{irr}(\alpha, F) = \mathrm{irr}(\beta, F)$.

Example 5.1.1. If $a, b \in \mathbb{R}$ with $b \neq 0$, the conjugate complex numbers $a + bi$ and $a - bi$ are both zeros of $x^2 - 2ax + a^2 + b^2$, which is irreducible in $\mathbb{R}[x]$.

Theorem 5.1.2. Let $\alpha, \beta \in \overline{F}$ be algebraic over F with $\deg(\alpha, F) = n$. The map $\psi_{\alpha,\beta} : F(\alpha) \to F(\beta)$ defined by

$$\psi_{\alpha,\beta}(c_0 + c_1\alpha + \cdots + c_{n-1}\alpha^{n-1}) = c_0 + c_1\beta + \cdots + c_{n-1}\beta^{n-1}, \forall c_i \in F$$

is an isomorphism (called **conjugation isomorphism**) if and only if α and β are conjugate over F.

Proof. ($\Rightarrow$). Let $\mathrm{irr}(\alpha, F) = a_0 + a_1 x + \cdots + a_n x^n$. Then $a_0 + a_1\alpha + \cdots + a_n\alpha^n = 0$, so

$$\psi_{\alpha,\beta}(a_0 + a_1\alpha + \cdots + a_n\alpha^n) = a_0 + a_1\beta + \cdots + a_n\beta^n = 0.$$

This implies that $\mathrm{irr}(\beta, F) | \mathrm{irr}(\alpha, F)$. Since both polynomials are monic and irreducible, then $\mathrm{irr}(\alpha, F) = \mathrm{irr}(\beta, F)$, so α and β are conjugate over F.

($\Leftarrow$). Suppose $\mathrm{irr}(\alpha, F) = \mathrm{irr}(\beta, F) = p(x)$. Then both the evaluation homomorphisms $\phi_\alpha : F[x] \to F(\alpha)$ and $\phi_\beta : F[x] \to F(\beta)$ have the same kernel $\langle p(x) \rangle$.

$$F(\beta) \xleftarrow{\phi_\beta} F[x] \xrightarrow{\phi_\alpha} F(\alpha)$$

$$F(\beta) \xleftarrow{\psi_\beta} F[x]/\langle p(x) \rangle \xrightarrow{\psi_\alpha} F(\alpha)$$

We have the natural isomorphisms $\psi_\alpha : F[x] \to F(\alpha)$ and $\psi_\beta :$ $F[x] \to F(\beta)$. Let $\psi_{\alpha,\beta} = \psi_\beta \psi_\alpha^{-1}$. Clearly, $\psi_{\alpha,\beta}$ is an isomorphism mapping $F(\alpha)$ onto $F(\beta)$. For $c_0 + c_1\alpha + \cdots + c_{n-1}\alpha^{n-1} \in F(\alpha)$, we have

$$\psi_{\alpha,\beta}(c_0 + c_1\alpha + \cdots + c_{n-1}\alpha^{n-1}) = \psi_\beta \psi_\alpha^{-1}(c_0 + c_1\alpha + \cdots + c_{n-1}\alpha^{n-1})$$
$$= \psi_\beta((c_0 + c_1 x + \cdots + c_{n-1}x^{n-1}) + \langle p(x) \rangle) = c_0 + c_1\beta + \cdots + c_{n-1}\beta^{n-1}.$$

Thus $\psi_{\alpha,\beta}$ is the isomorphism defined in the statement of the theorem. $\square$

Corollary 5.1.3. *Let $\alpha \in \overline{F}$ be algebraic over F, and ψ be an isomorphism mapping $F(\alpha)$ onto a subfield of $\overline{F}$ such that $\psi(a) = a$ for $a \in F$. Then $\beta = \phi(\alpha)$ is a conjugate of α over F.*

Proof. Let $\mathrm{irr}(\alpha, F) = a_0 + a_1 x + \cdots + a_n x^n$. Then

$$a_0 + a_1\alpha + \cdots + a_n\alpha^n = 0,$$

$$0 = \psi(a_0 + a_1\alpha + \cdots + a_n\alpha^n) = a_0 + a_1\psi(\alpha) + \cdots + a_n\psi(\alpha)^n,$$

and $\beta = \psi(\alpha)$ is a conjugate of α. $\square$

A special case of the above corollary is a familiar result.

Corollary 5.1.4. *Let $f(x) \in \mathbb{R}[x]$. If $f(a + bi) = 0$ for $a + bi \in \mathbb{C}$, where $a, b \in R$, then $f(a - bi) = 0$ also.*

Proof. We have seen that $\mathbb{C} = \mathbb{R}(i)$. Now $\mathrm{irr}(i, \mathbb{R}) = \mathrm{irr}(-i, \mathbb{R}) = x^2 + 1$, so i and $-i$ are conjugate over $\mathbb{R}$. By the previous Theorem, the conjugation map $\psi = \psi_{i,-i} : \mathbb{C} \to \mathbb{C}$ where $\psi(a + bi) = a - bi$ is an isomorphism. Since $f(x) \in \mathbb{R}[x]$, applying ψ to $f(a + bi) = 0$ we obtain that $f(a - bi) = 0$. $\square$

Example 5.1.2. *Consider $\mathbb{Q}(\sqrt{2})$ over $\mathbb{Q}$. The zeros of $\mathrm{irr}(\sqrt{2}, \mathbb{Q}) = x^2 - 2$ are $\pm\sqrt{2}$, so they are conjugate over $\mathbb{Q}$. The conjugation isomorphism $\psi_{\sqrt{2},-\sqrt{2}} : \mathbb{Q}(\sqrt{2}) \to \mathbb{Q}(\sqrt{2})$ defined by*

$$\psi_{\sqrt{2},-\sqrt{2}}(a + b\sqrt{2}) = a - b\sqrt{2}$$

is an automorphism of $\mathbb{Q}(\sqrt{2})$.

As illustrated in the preceding corollary and example, a field may have a nontrivial isomorphism onto itself.

Definition 5.1.5. *An isomorphism of a field F onto itself is an* **automorphism of the field** *of F. We denote the set of all automorphisms of F by $\mathrm{Aut}(F)$.*

Definition 5.1.6. *Let $E \leq K$ be fields, σ an isomorphism of E onto a subfield of K. Then $a \in E$ is* **left fixed by** σ *if $\sigma(a) = a$. A collection S of isomorphisms of E* **leaves a subfield** F **of E fixed** *if*

$$sigma(a) = a, \forall a \in F, \sigma \in S.$$

If $\{\sigma\}$ leaves F fixed, then σ **leaves F fixed.**

Example 5.1.3. *Let $E = \mathbb{Q}(\sqrt{2}, \sqrt{3})$. The map $\sigma : E \to E$ defined by*

$$\sigma(a + b\sqrt{2} + c\sqrt{3} + d\sqrt{6}) = a + b\sqrt{2} - c\sqrt{3} - d\sqrt{6}$$

for $a, b, c, d \in \mathbb{Q}$ is an automorphism of E; it is the conjugation isomorphism $\psi_{\sqrt{3}, -\sqrt{3}}$ of E onto itself if we view E as $(\mathbb{Q}(\sqrt{2}))(\sqrt{3})$. We see that σ leaves $\mathbb{Q}(\sqrt{2})$ fixed.

Theorem 5.1.7. *Let $H = \{\sigma_k | k \in I\} \subseteq \mathrm{Aut}(E)$ where E is a field. Then the set $E_H = \{a \in E : \sigma_k(a) = a \,\forall k \in I\}$ is a subfield of E.*

Proof. Let $a, b \in E_H$, i.e., $\sigma_k(a) = a$ and $\sigma_k(b) = b$ for all $k \in I$. Then

$$\sigma_k(a \pm b) = \sigma_k(a) \pm \sigma_k(b) = a \pm b,$$

$$\sigma_k(ab) = \sigma_k(a)\sigma_k(b) = ab, \forall k \in I,$$

i.e., $a \pm b, ab \in E_H$. Also, if $b \neq 0$, then

$$\sigma_k(a/b) = \sigma_k(a)/\sigma_k(b) = a/b$$

for all $k \in I$, i.e., $a/b \in E_H$. Since the σ_k are automorphisms, $\sigma_k(0) = 0$, $\sigma_k(1) = 1$ for all $k \in I$, i.e., $0, 1 \in E_H$. Thus $E_H \leq E$. $\square$

Definition 5.1.8. *The field E_H in the above Theorem is the* **fixed field** *of H. For a single automorphism σ, we shall refer to E_σ as the* **fixed field** *of σ.*

Example 5.1.4. *Consider the conjugation automorphism $\psi_{\sqrt{2}, -\sqrt{2}}$ of $\mathbb{Q}(\sqrt{2})$ given in the previous example. For $a, b \in \mathbb{Q}$, we have*

$$\psi_{\sqrt{2}, -\sqrt{2}}(a + b\sqrt{2}) = a - b\sqrt{2},$$

and $a - b\sqrt{2} = a + b\sqrt{2}$ if and only if $b = 0$. Thus the fixed field of $\psi_{\sqrt{2}, -\sqrt{2}}$ is $\mathbb{Q}$.

Theorem 5.1.9. *The set $\mathrm{Aut}(E)$ is a subgroup of the symmetric group $(S_E, \circ)$.*

Proof. We know that $(S_E, \circ)$ is a group, where the multiplication is map composition. The identity map $\iota : E \to E$ is in $\mathrm{Aut}(E)$. For $\sigma, \tau \in \mathrm{Aut}(E)$, it is easy to see that $\sigma\tau, \sigma^{-1} \in \mathrm{Aut}(E)$. Thus $\mathrm{Aut}(E)$ is a subgroup of S_E.   □

Theorem 5.1.10. *Let E be a field, and let F be a subfield of E. Then the set*

$$\mathrm{Gal}(E/F) = \{\sigma \in \mathrm{Aut}(E) : \sigma(a) = a \ \forall a \in F\}$$

is a subgroup of $\mathrm{Aut}(E)$. Furthermore, $F \leq E_{\mathrm{Gal}(E/F)}$.

Proof. For $\sigma, \tau \in \mathrm{Gal}(E/F)$ and $a \in F$, we have

$$(\sigma\tau)(a) = \sigma(\tau(a)) = \sigma(a) = a,$$

so $\sigma\tau \in \mathrm{Gal}(E/F)$. Of course, the identity automorphism ι is in $\mathrm{Gal}(E/F)$. Also, if $\sigma(a) = a$ for $a \in F$, then $a = \sigma^{-1}(a)$ so $\sigma \in \mathrm{Gal}(E/F)$ implies that $\sigma^{-1} \in \mathrm{Gal}(E/F)$. Thus $\mathrm{Gal}(E/F)$ is a subgroup of $\mathrm{Aut}(E)$.

By definition we know that $\sigma(a) = a$ for $a \in F$ and $\sigma \in \mathrm{Gal}(E/F)$. So $F \leq E_{\mathrm{Gal}(E/F)}$.   □

Definition 5.1.11. *The group $\mathrm{Gal}(E/F)$ of the preceding theorem is called the* **Galois group of E over F.**

Example 5.1.5. *Find $\mathrm{Gal}(\mathbb{Q}(\sqrt{2}, \sqrt{3}), \mathbb{Q})$.*

Solution. We know that $[\mathbb{Q}(\sqrt{2}, \sqrt{3}) : \mathbb{Q}] = 4$. If we view $\mathbb{Q}(\sqrt{2}, \sqrt{3})$ as $(\mathbb{Q}(\sqrt{3}))(\sqrt{2})$, the conjugation isomorphism $\psi_{\sqrt{2},-\sqrt{2}}$ defined by

$$\psi_{\sqrt{2},-\sqrt{2}}(a + b\sqrt{2}) = a - b\sqrt{2}$$

for $a, b \in \mathbb{Q}(\sqrt{3})$ is an automorphism of $\mathbb{Q}(\sqrt{2}, \sqrt{3})$ having $\mathbb{Q}(\sqrt{3})$ as fixed field. Similarly, we have the automorphism $\psi_{\sqrt{3},-\sqrt{3}}$ of $\mathbb{Q}(\sqrt{2}, \sqrt{3})$ having $\mathbb{Q}(\sqrt{2})$ as fixed field. The automorphisms $\psi_{\sqrt{2},-\sqrt{2}}\psi_{\sqrt{3},-\sqrt{3}}$ moves both $\sqrt{2}$ and $\sqrt{3}$, that is, leaves neither number fixed. Let

$$\iota = \text{ the identity automorphism,}$$

$$\sigma_1 = \psi_{\sqrt{2},-\sqrt{2}},$$

$$\sigma_2 = \psi_{\sqrt{3},-\sqrt{3}},$$

$$\sigma_3 = \psi_{\sqrt{2},-\sqrt{2}}\psi_{\sqrt{3},-\sqrt{3}}.$$

The group of all automorphisms of $\mathbb{Q}(\sqrt{2}, \sqrt{3})$ has a fixed field. This fixed field must contain $\mathbb{Q}$, since every automorphism of a field leaves 1 and hence the prime subfield fixed. A basis for $\mathbb{Q}(\sqrt{2}, \sqrt{3})$ over $\mathbb{Q}$ is $\{1, \sqrt{2}, \sqrt{3}, \sqrt{6}\}$. Since $\sigma_1(\sqrt{2}) = -\sqrt{2}, \sigma_1(\sqrt{6}) = -\sqrt{6}$ and $\sigma_2(\sqrt{3}) = -\sqrt{3}$, we see that $\mathbb{Q}$ is exactly the fixed field of $\{\iota, \sigma_1, \sigma_2, \sigma_3\}$. It is readily checked that $G = \{\iota, \sigma_1, \sigma_2, \sigma_3\}$ is a group under automorphism multiplication (function composition). For example,

$$\sigma_1\sigma_3 = \psi_{\sqrt{2},-\sqrt{2}}(\psi_{\sqrt{2},-\sqrt{2}}\psi_{\sqrt{3},-\sqrt{3}}) = \psi_{\sqrt{3},-\sqrt{3}} = \sigma_2.$$

The group G is actually isomorphic to $(\mathbb{Z}_2, +) \times (\mathbb{Z}_2, +)$. We can show that $G = \mathrm{Gal}(\mathbb{Q}(\sqrt{2}, \sqrt{3})/\mathbb{Q})$, because every automorphism τ of $\mathbb{Q}(\sqrt{2}, \sqrt{3})$ maps $\sqrt{2}$ onto either $\pm\sqrt{2}$. Similarly, τ maps $\sqrt{3}$ onto either $\pm\sqrt{3}$. But since $1, \sqrt{2}, \sqrt{3}, \sqrt{2}\sqrt{3}$ is a basis for $\mathbb{Q}(\sqrt{2}, \sqrt{3})$ over $\mathbb{Q}$, an automorphism of $\mathbb{Q}(\sqrt{2}, \sqrt{3})$ leaving $\mathbb{Q}$ fixed is determined by its values on $\sqrt{2}$ and $\sqrt{3}$. Now, $\iota, \sigma_1, \sigma_2,$ and σ_3 give all possible combinations of values on $\sqrt{2}$ and $\sqrt{3}$, and hence are all possible automorphisms of $\mathbb{Q}(\sqrt{2}, \sqrt{3})$.

Note that $\mathrm{Gal}(\mathbb{Q}(\sqrt{2}, \sqrt{3})/\mathbb{Q})$ has order 4, and $[\mathbb{Q}(\sqrt{2}, \sqrt{3}) : \mathbb{Q}] = 4$. This holds for a general situation, as we shall prove later.   □

For any finite field F we shall show later that the group $\mathrm{Aut}(F)$ is cyclic. Actually the group $\mathrm{Aut}(F)$ has a canonical generator, the Frobenius automorphism given by the next theorem.

Theorem 5.1.12. *Let F be a finite field of characteristic p. Then the map $\sigma_p : F \to F$ defined by $\sigma_p(a) = a^p$ for $a \in F$ is an automorphism, the* **Frobenius automorphism**, *called of F. Also, $F_{\sigma_p} \simeq \mathbb{Z}_p$.*

Proof. Let $a, b \in F$. We see that $(a + b)^p = a^p + b^p$. So

$$\sigma_p(a + b) = (a + b)^p = a^p + b^p = \sigma_p(a) + \sigma_p(b).$$

$$\sigma_p(ab) = (ab)^p = a^p b^p = \sigma_p(a)\sigma_p(b),$$

i.e., σ_p is at least a homomorphism. If $\sigma_p(a) = 0$, then $a^p = 0$, and $a = 0$, i.e., $\ker(\sigma_p) = \{0\}$. Then σ_p is a one-to-one map. Since F is finite, σ_p is also onto. Thus σ_p is an automorphism of F.

From Theorem 4.4.5 we know that $\mathbb{Z}_p = \{a \in F : a^p = a\}$, and $F_{\sigma_p} = \{a \in \overline{F} : \sigma_p(a) = a\}$. We see that $\mathbb{Z}_p = F_{\sigma_p}$.   □

5.2. The isomorphism extension theorem.

Remember that we always assume that E, F are fields, and $\overline{F}$ is an algebraic closure of F.

Theorem 5.2.1 (Isomorphism Extension Theorem). *Let E be an algebraic extension of a field F. Let σ be an isomorphism of F onto a field F'. Then σ can be extended to an isomorphism τ of E onto a subfield of $\overline{F'}$, i.e., $\tau|_F = \sigma$.*

$$
\begin{array}{ccc}
F & \xrightarrow{\sigma} & F' \\
\cap & & \cap \\
E & \xrightarrow{\exists \tau} & \overline{F'}
\end{array}
$$

Proof. Let $\mathcal{S}$ be the set of all pairs (L, λ), where L is a field such that $F \leq L \leq E$ and λ is an isomorphism of L onto a subfield of $\overline{F'}$ such that $\lambda(a) = \sigma(a)$ for $a \in F$. Clearly, $(F, \sigma) \in \mathcal{S}$. Define a partial ordering on $\mathcal{S}$ by $(L_1, \lambda_1) \leq (L_2, \lambda_2)$, if $L_1 \leq L_2$ and $\lambda_1(a) = \lambda_2(a)$ for $a \in L_1$. This relation $\leq$ gives a partial ordering of $\mathcal{S}$.

Let $\mathcal{T} = \{(H_i, \lambda_i) : i \in I\}$ be a chain in $\mathcal{S}$. We claim that $H = \cup_{i \in I} H_i$ is a subfield of E. Let $a, b \in H$, where $a \in H_1$ and $b \in H_2$; then either $H_1 \leq H_2$ or $H_2 \leq H_1$, since $\mathcal{T}$ is a chain. Say, $H_1 \leq H_2$. Then $a, b \in H_2$, so $a \pm b, ab$, and a/b for $b \neq 0$ are all in H_2 and hence in H. Since for each $i \in I, F \subset H_i \subset E$, we have $F \subset H \subset E$. Thus $H \leq E$.

Define $\lambda : H \to \overline{F'}$ as follows. Let $c \in H$. Then $c \in H_i$ for some $i \in I$, and let $\lambda(c) = \lambda_i(c)$. The map λ is well defined because if $c \in H_1$ and $c \in H_2$, then either $(H_1, \lambda_1) \leq (H_2, \lambda_2)$ or $(H_2, \lambda_2) \leq (H_1, \lambda_1)$, since $\mathcal{T}$ is a chain. In either case, $\lambda_1(c) = \lambda_2(c)$. We claim that λ is an isomorphism of H onto a subfield of F'. If $a, b \in H$ then there is an H_i such that $a, b \in H_i$, and

$$\lambda(a + b) = \lambda_i(a + b) = \lambda_i(a) + \lambda_i(b) = \lambda(a) + \lambda(b).$$

Similarly,

$$\lambda(ab) = \lambda_i(ab) = \lambda_i(a)\lambda_i(b) = \lambda(a)\lambda(b).$$

If $\lambda(a) = 0$, then $a \in H_i$ for some i implies that $\lambda_i(a) = 0$, so $a = 0$. Therefore, λ is an isomorphism. Thus $(H, \lambda) \in \mathcal{S}$, and it is clear from our definitions of H and λ that (H, λ) is an upper bound for $\mathcal{T}$.

We have shown that every chain of $\mathcal{S}$ has an upper bound in $\mathcal{S}$. So the hypotheses of Kuratowski-Zorn lemma are satisfied. Hence there exists a maximal element $(K, \tau) \in \mathcal{S}$. Let $\tau(K) = K'$, where

$K' \leq \overline{F'}$. If $K \neq E$, take $\alpha \in E \setminus K$. Now α is algebraic over F, so α is algebraic over K. Also, let $p(x) = \text{irr}(\alpha, K)$. Extending the isomorphism τ we have the ring isomorphism $\tau_x : K[x] \to K'[x]$ by

$$\tau_x(a_0 + a_1 x + \cdots + a_n x^n) = \tau(a_0) + \tau(a_1)x + \cdots + \tau(a_n)x^n, \forall a_i \in K.$$

Let $q(x) = \tau_x(p(x))$. Since τ_x is an isomorphism, $q(x)$ is irreducible in $K'[x]$, and we have the induced isomorphism

$$\overline{\tau} : K[x]/\langle p(x) \rangle \rightarrowtail K'[x]/\langle q(x) \rangle$$

from τ_x. Let ψ_α be the canonical isomorphism

$$\psi_\alpha : K[x]/\langle p(x) \rangle \rightarrowtail K(\alpha),$$

corresponding to the evaluation homomorphism $\phi_\alpha : K[x] \to K(\alpha)$. Since $K' \leq \overline{F'}$, there is a zero α' of $q(x)$ in $\overline{F'}$. Let

$$\psi_{\alpha'} : K'[x]/\langle q(x) \rangle \rightarrowtail K'(\alpha')$$

be the isomorphism analogous to ψ_α.

$$K(\alpha) \xleftarrow{\phi_\alpha} K[x] \xrightarrow{\tau_x} K'[x] \xrightarrow{\phi_{\alpha'}} K'(\alpha')$$

$$K(\alpha) \xleftarrow{\psi_\alpha} K[x]/\langle p(x) \rangle \xrightarrow{\overline{\tau}} K'[x]/\langle q(x) \rangle \xrightarrow{\psi_{\alpha'}} K'(\alpha')$$

Then the composition of maps

$$\psi_{\alpha'}\overline{\tau}\psi_\alpha^{-1} : K(\alpha) \to K'(\alpha')$$

is an isomorphism of $K(\alpha)$ onto a subfield of $\overline{F'}$. Since

$$\psi_{\alpha'}\overline{\tau}\psi_\alpha^{-1}(k) = \psi_{\alpha'}\overline{\tau}(k) = \psi_{\alpha'}\tau(k) = \tau(k), \forall k \in K,$$

so, $(K, \tau) < (K(\alpha), \psi_{\alpha'}\overline{\tau}\psi_\alpha^{-1})$ which contradicts that (K, τ) is maximal. Therefore we must have had $K = E$. $\qquad\square$

Example 5.2.1. *We have the automorphism $\psi_{\sqrt{2}, -\sqrt{2}} : \mathbb{Q}(\sqrt{2}) \to \mathbb{Q}(\sqrt{2})$. Can we extend $\psi_{\sqrt{2}, -\sqrt{2}}$ to an automorphism of $\mathbb{R}$? (No. See Example 1.2.1.) Can we extend $\psi_{\sqrt{2}, -\sqrt{2}}$ to an automorphism of $\mathbb{C}$? (Yes, a lot. But at this moment we cannot prove this.)*

We give as a corollary the existence of an extension of one of our conjugation isomorphisms $\psi_{\alpha, \beta}$, as discussed at the beginning of this section.

Corollary 5.2.2. *If $F \leq E \leq \overline{F}$ where E is algebraic over F, and $\alpha, \beta \in E$ are conjugate over F, then the conjugation isomorphism $\psi_{\alpha,\beta} : F(\alpha) \to F(\beta)$ can be extended to an isomorphism of E onto a subfield of $\overline{F}$.*

Proof. Proof of this corollary is immediate from Theorem 5.2.1 if in the statement of the theorem we replace F by $F(\alpha), F'$ by $F(\beta)$, and $\overline{F'}$ by $\overline{F}$. □

Corollary 5.2.3. *Let $\sigma : F \to F'$ be an isomorphism of the fields. Then σ can be extended to an isomorphism $\overline{\sigma} : \overline{F} \to \overline{F'}$.*

Proof. By Theorem 5.2.1, σ can be extended to an isomorphism $\tau : \overline{F}$ onto a subfield of $\overline{F'}$. We need only show that τ is onto $\overline{F'}$. But the map $\tau^{-1} : \tau(\overline{F}) \to \overline{F}$ can be extended to an isomorphism of $\overline{F'}$ onto a subfield of $\overline{F}$. Since τ^{-1} is already onto $\overline{F}$, we must have $\tau(\overline{F}) = \overline{F'}$. □

As a direct consequence of this corollary, we see that an algebraic closure of F is unique, up to an isomorphism over F, i.e.,

Corollary 5.2.4. *Let $\overline{F}$ and $\overline{F}'$ be two algebraic closures of F. Then there is an isomorphism $\tau : \overline{F} \to \overline{F'}$ such that $\tau(a) = a$ for any $a \in F$.*

Theorem 5.2.5. *Let E be a finite extension of a field F, and $\sigma : F \to F'$ be an isomorphism of fields. Then the number of extensions of σ to an isomorphism τ of E onto a subfield of $\overline{F'}$ is finite, and independent of $F', \overline{F'}$, and σ. That is, the number of extensions is completely determined by the two fields E and F.*

Proof. By Corollary 5.2.3 we can extend $\sigma^{-1} : F' \to F$ to an isomorphism $\overline{\sigma} : \overline{F'} \to \overline{F}$. Each isomorphism $\tilde{\sigma}$ of E onto a subfield of $\overline{F'}$ that extends σ, one-to-one corresponds an isomorphism $\overline{\sigma}\tilde{\sigma} : E \to \overline{F}$. Indeed, if isomorphisms $\tilde{\sigma}, \tilde{\sigma}'$ of E onto a subfield of $\overline{F'}$ extends σ, then $\overline{\sigma}\tilde{\sigma} = \overline{\sigma}\tilde{\sigma}'$ iff $\tilde{\sigma} = \tilde{\sigma}'$.

$$
\begin{array}{ccccc}
F & \overset{\sigma}{\rightarrowtail} & F' & \overset{\sigma^{-1}}{\rightarrowtail} & F \\
\cap & & \cap & & \cap \\
E & \overset{?\exists\tilde{\sigma}}{\longrightarrow} & \overline{F'} & \overset{\overline{\sigma}}{\rightarrowtail} & \overline{F}
\end{array}
\qquad
\begin{array}{ccc}
F & = & F \\
\cap & & \cap \\
E & \overset{?\exists\tau}{\longrightarrow} & \overline{F}
\end{array}
$$

In order to prove the theorem we may assume that $F' = F$ and $\sigma = \mathrm{id}_F$. Since E is a finite extension of F, we take a basis of E over F : $\beta = \{a_1, a_2, \cdots, a_n\}$. Let $p_i(x) = \mathrm{irr}(a_i, F)$. Let $\tau : E \to \overline{F}$ be any isomorphism from E onto a subfield of $\overline{F}$ with $\tau(a) = a$ for any $a \in F$. We know that τ is uniquely determined by $\tau(a_1), \tau(a_2), \cdots, \tau(a_n) \in \overline{F}$.

Applying τ to $p_i(a_i) = 0$, we see that $p_i(\tau(a_i)) = 0$, that is a_i and $\tau(a_i)$ are zeros of $p_i(x)$. Each $p_i(x)$ has only finitely many zeros. We conclude that we have only finitely may τ. This completes the proof. $\square$

Definition 5.2.6. *Let E be a finite extension of a field F. Let $S(E/F)$ be the set of all isomorphisms of E onto a subfield of $\overline{F}$ over F. Then we call $|S(E/F)|$ the* **index** *of E over F, denoted by $\{E : F\}$.*

You may compare the two sets $\mathrm{Gal}(E/F)$ and $S(E/F)$.

Corollary 5.2.7. *If $F \leq E \leq K$, where K is a finite extension field of F, then $\{K : F\} = \{K : E\}\{E : F\}$.*

Proof.
$$
\begin{array}{ccccc}
F & \leq & E & \leq & K \\
\| & & \sigma \downarrow & & \tau \downarrow \\
F & \leq & \overline{F} & = & \overline{F}
\end{array}
$$
By Theorem 5.2.1, each of the $\{E : F\}$ isomorphisms σ of E onto a subfield of $\overline{F}$ over F has $\{K : E\}$ extensions to an isomorphism τ of K onto a subfield of $\overline{F}$. $\square$

Example 5.2.2. *Consider $E = \mathbb{Q}(\sqrt{2}, \sqrt{3})$ over $\mathbb{Q}$. We know that $\{E : \mathbb{Q}\} = [E : \mathbb{Q}] = 4$. Also, $\{E : \mathbb{Q}(\sqrt{2})\} = 2$, and $\{\mathbb{Q}(\sqrt{2}) : \mathbb{Q}\} = 2$, so*
$$
4 = \{E : \mathbb{Q}\} = \{E : \mathbb{Q}(\sqrt{2})\}\{\mathbb{Q}(\sqrt{2}) : \mathbb{Q}\} = (2)(2).
$$

5.3. Splitting fields.

We will determine for what extension field $F \leq E \leq \overline{F}$, every isomorphic mapping of E onto a subfield of $\overline{F}$ over F is actually an automorphism of E.

Definition 5.3.1. *Let F be a field with algebraic closure $\overline{F}$. Let $P = \{f_i(x) : i \in I\}$ be a collection of polynomials in $F[x]$. A field $E \leq \overline{F}$ is the* **splitting field** *of P* **over** *F if E is the smallest subfield of $\overline{F}$ containing F and all the zeros in $\overline{F}$ of each of $f_i(x) \in P$. A field $K \leq \overline{F}$ is a* **splitting field over** *F if it is the splitting field of some set of polynomials in $F[x]$.*

Example 5.3.1. *We see that* $\mathbb{Q}[\sqrt{2}, \sqrt{3}]$ *is a splitting field of* $\{x^2 - 2, x^2 - 3\}$ *and also of* $x^4 - 5x^2 + 6$.

For one polynomial $f(x) \in F[x]$, we shall often refer to the splitting field of $\{f(x)\}$ over F as the **splitting field of** $f(x)$ **over** F, denoted by $F_{f(x)}$.

Let $P = \{f_i(x) : i \in I\} \subset F[x]$, and let

$$\mathcal{R}_P = \{a \in \overline{F} : f_i(a) = 0 \text{ for some } i\}.$$

Then we can see that $F(\mathcal{R}_P)$ is the splitting field of $\{f_i(x) : i \in I\}$ over F. If $P = \{f(x)\}$ we will simple denote $\mathcal{R}_{f(x)} = \mathcal{R}_{\{f(x)\}}$.

Theorem 5.3.2. *A field* E, *where* $F \leq E \leq \overline{F}$, *is a splitting field over* F *if and only if every* $\tau \in \mathrm{Gal}(\overline{F}/F)$ *maps* E *onto itself, i.e.,* $\tau|_E \in \mathrm{Gal}(E/F)$.

Proof. ($\Rightarrow$). Let E be a splitting field over F in $\overline{F}$ of $P = \{f_i(x) : i \in I\}$. Let $\mathcal{R}_P$ be defined above. Then $E = F(\mathcal{R}_P)$. For any $\alpha \in \overline{F}$, $f_i(\alpha) = 0$ if and only if $f_i(\tau(\alpha)) = 0$. So $\tau(\mathcal{R}_P) = \mathcal{R}_P$. Thus $\tau(E) = \tau(F(\mathcal{R}_P)) = F(\tau(\mathcal{R}_P)) = F(\mathcal{R}_P) = E$, i.e., $\tau|_E \in \mathrm{Gal}(E/F)$.

($\Leftarrow$). Suppose that $\tau|_E \in \mathrm{Gal}(E/F)$ for every $\tau \in \mathrm{Gal}(\overline{F}/F)$. Take a basis β for E over F. For each $a \in \beta$ let $p_a(x) = \mathrm{irr}(a, F)$. Let $P = \{p_a(x) : a \in \beta\}$. We claim that $E = F(\mathcal{R}_p)$, i.e., E is the splitting field over F in $\overline{F}$ of P. Since $E \subseteq F(\mathcal{R}_p)$, it is enough to show that $F(\mathcal{R}_p) \subseteq E$.

Take any $g(x) = p_a(x) \in P$. If b is any zero of $g(x)$ in $\overline{F}$, then there is a conjugation isomorphism $\psi_{a,b}$ of $F(a)$ onto $F(b)$ over F. By Theorem 4.2.1, $\psi_{a,b}$ can be extended to an automorphism τ of $\overline{F}$. Since $\tau(E) = E$ we see that $\tau(a) = b \in E$. Consequently, $\mathcal{R}_P \in E$. Thus $F(\mathcal{R}_p) \subseteq E$. $\qquad\square$

Definition 5.3.3. *Let* E *be an extension field of a field* F. *A polynomial* $f(x) \in F[x]$ **splits** *in* E *if it factors into a product of linear factors in* $E[x]$.

Corollary 5.3.4. *If* $E \leq \overline{F}$ *is a splitting field over* F, *then every irreducible polynomial in* $F[x]$ *having a zero in* E *splits in* E.

Proof. Since E is a splitting field over F in $\overline{F}$, then $\tau|_E \in \mathrm{Gal}(E/F)$ for every $\tau \in \mathrm{Gal}(\overline{F}/F)$. The last paragraph of the proof of above Theorem showed precisely that any irreducible polynomial

$g(x) \in F[x]$ having a zero in E have all zeros in E, i.e., its factorization into linear factors in $\overline{F}[x]$, actually takes place in $E[x]$, so $g(x)$ splits in E. $\square$

Corollary 5.3.5. *If $E \leq \overline{F}$ is a splitting field over F, then $S(E/F) =$ Gal(E/F). If E is further of finite degree over F, then $\{E : F\} = |\text{Gal}(E/F)|$.*

Proof. It is clear that $\text{Gal}(E/F) \subseteq S(E/F)$. Each $\sigma \in S(E/F)$ can be extended to an automorphism τ of $\overline{F}$.

$$
\begin{array}{ccccc}
F & \leq & E & \leq & \overline{F} \\
\| & & \sigma \downarrow & & \tau \downarrow \\
F & \leq & \overline{F} & = & \overline{F}
\end{array}
$$

Since E is a splitting field over F, then, $\sigma = \tau|_E \in \text{Gal}(E/F)$. Thus $S(E/F) = \text{Gal}(E/F)$.

The equation $\{E : F\} = |\text{Gal}(E/F)|$ then follows clearly. $\square$

Corollary 5.3.6. *Let $\alpha \in \overline{F}$ be algebraic over F. Then $\{F(\alpha) : F\}$ is the number of different zeros of $\text{irr}(\alpha, F)$.*

Proof. By Theorem 5.1.2 we know that $\alpha, \sigma(\alpha)$ are conjugate over F for each $\sigma \in S(F(\alpha)/F)$, and each conjugate β of α can have a conjugation isomorphism $\psi_{\alpha,\beta}$. $\square$

If $F \leq E \leq K$ is a chain of field extensions such that $F \leq E$ is splitting and $E \leq K$ is splitting, it is false to conclude that $F \leq K$ is splitting. For example $\mathbb{Q} \leq \mathbb{Q}[\sqrt{2}] \leq \mathbb{Q}[\sqrt{\sqrt{2} - 1}]$.

In the next section we will determine conditions under which $|\text{Gal}(E/F)| = \{E : F\} = [E : F]$ for finite extensions E of F.

Example 5.3.2. *Find the splitting field E of $x^3 - 2$ over $\mathbb{Q}$.*

Solution. We know that $x^3 - 2$ does not split in $\mathbb{Q}(\sqrt[3]{2})$, for $\mathbb{Q}(\sqrt[3]{2}) < \mathbb{R}$ and only one zero of $x^3 - 2$ is real. Thus $x^3 - 2$ factors in $(\mathbb{Q}(\sqrt[3]{2}))[x]$ into a linear factor $x - \sqrt[3]{2}$ and an irreducible quadratic factor. So $[E : \mathbb{Q}(\sqrt[3]{2})] = 2$. Then

$$[E : \mathbb{Q}] = [E : \mathbb{Q}(\sqrt[3]{2})][\mathbb{Q}(\sqrt[3]{2}) : \mathbb{Q}] = (2)(3) = 6.$$

We can easily see that

$$\frac{-1 + i\sqrt{3}}{2}\sqrt[3]{2} \text{ and } \frac{-1 - i\sqrt{3}}{2}\sqrt[3]{2}$$

are the other zeros of $x^3 - 2$ in $\mathbb{C}$. Thus $E = \mathbb{Q}(\sqrt[3]{2}, i\sqrt{3})$. (This is not the same field as $\mathbb{Q}(\sqrt[3]{2}, i, \sqrt{3})$, which is of degree 12 over $\mathbb{Q}$.) $\square$

5.4. Separable extensions.

We now assume that all algebraic extensions of a field F under consideration are contained in one fixed algebraic closure $\overline{F}$ of F.

Our aim in this section is to determine, for a finite extension E of F, what conditions ensure $\{E : F\} = [E : F]$. The key to answering this question is to consider the multiplicity of zeros of polynomials.

Definition 5.4.1. *Let* $f(x) \in F[x]$. *An element* $\alpha \in \overline{F}$ *such that* $f(\alpha) = 0$ *is a* **zero** *of* $f(x)$ *of* **multiplicity** r *if* $(x - \alpha)^r | f(x)$ *but* $(x - \alpha)^{r+1} \nmid f(x)$ *in* $\overline{F}[x]$.

Theorem 5.4.2. *Let* $f(x)$ *be irreducible in* $F[x]$. *Then all zeros of* $f(x)$ *in* $\overline{F}$ *have the same multiplicity.*

Proof. Let $\alpha, \beta \in \mathcal{R}_{f(x)}$. We have a conjugation isomorphism $\psi_{\alpha,\beta} : F(\alpha) \to F(\beta)$. Then $\psi_{\alpha,\beta}$ can be extended to an isomorphism $\tau : \overline{F} \to \overline{F}$. Thus τ induces a natural isomorphism $\tau_x : \overline{F}[x] \to \overline{F}[x]$, with $\tau_x(x) = x$. Now τ_x leaves $f(x)$ fixed, since $f(x) \in F[x]$ and $\psi_{\alpha,\beta}$ leaves F fixed. However,

$$\tau_x((x - \alpha)^r) = (x - \beta)^r,$$

which shows that the multiplicity of β in $f(x)$ is greater than or equal to the multiplicity of α. A symmetric argument gives the reverse inequality, so the multiplicity of α equals that of β. □

Corollary 5.4.3. *If* $f(x) \in F[x]$ *is irreducible, then* $f(x)$ *has a factorization in* $\overline{F}[x]$ *of the form*

$$a \prod_{k=1}^{n} (x - \alpha_k)^r,$$

where the α_k *are the distinct zeros of* $f(x)$ *in* $\overline{F}$ *and* $a \in F$.

Proof. The corollary is immediate from the previous Theorem. □

To illustrate the above result let us work out some examples next.

Example 5.4.1. *Let* $E = \mathbb{Z}_p(y)$, *where* y *is an indeterminate. Let* $\alpha = y^p$, *and let* $F = \mathbb{Z}_p(\alpha) \leq E$. *Now* $E = F(y)$ *is algebraic over* F, *since* y *is a zero of* $(x^p - \alpha) \in F[x]$. *Then* $\mathrm{irr}(y, F) | x^p - \alpha$ *in* $F[x]$.

Since $F(y) \neq F$, *we must have the degree of* $\mathrm{irr}(y, F) \geq 2$. *Note that* $x^p - \alpha = x^p - y^p = (x - y)^p$, *since* E *has characteristic* p. *Thus* $\mathrm{irr}(y, F) | (x - y)^p$, *and* $\deg(y, F) = k > 1$. *Then* $\mathrm{irr}(y, F) = (x - y)^k \in$

$F[x]$. We deduce that $k = p$ and $\mathrm{irr}(y, F) = x^p - \alpha$, so the multiplicity of y is p.

Theorem 5.4.4. *If E is a finite extension of F, then $\{E : F\} \big| [E : F]$.*

Proof. If E is finite over F, then $E = F(\alpha_1, \ldots, \alpha_n)$, for some $\alpha_k \in \overline{F}$. Let $\mathrm{irr}(\alpha_k, F(\alpha_1, \ldots, \alpha_{k-1}))$ have α_k as one of n_k distinct zeros that are all of a common multiplicity r_k. From Corollary 5.3.6 we know that

$$[F(\alpha_1, \ldots, \alpha_k) : F(\alpha_1, \ldots, \alpha_{k-1})] = n_k r_k$$

$$= \{F(\alpha_1, \ldots, \alpha_k) : F(\alpha_1, \ldots, \alpha_{k-1})\} r_k.$$

From Corollaries 4.1.5 and 5.2.7 we see that

$$[E : F] = \prod_k n_k r_k, \quad \text{and} \quad \{E : F\} = \prod_k n_k.$$

Therefore, $\{E : F\} \big| [E : F]$. $\qquad\square$

Definition 5.4.5. *A finite extension E of F is a **separable extension** of F if $\{E : F\} = [E : F]$. An algebraic element α of $\overline{F}$ is separable over F if $F(\alpha)$ is a **separable extension** of F. An irreducible polynomial $f(x) \in F[x]$ is **separable over** F if every zero of $f(x)$ in $\overline{F}$ is separable over F.*

Example 5.4.2. (a). *The field $E = \mathbb{Q}[\sqrt{2}, \sqrt{3}]$ is separable over $\mathbb{Q}$ since we saw in previous examples that $\{E : \mathbb{Q}\} = 4 = [E : \mathbb{Q}]$.*

(b). *The extension E is not a separable extension of F in Example 5.4.1.*

By Corollary 5.1.3 we know that $\{F(\alpha) : F\}$ is the number of distinct zeros of $\mathrm{irr}(\alpha, F)$. Also, the multiplicity of α in $\mathrm{irr}(\alpha, F)$ is the same as the multiplicity of each conjugate of α over F, by Theorem 5.4.2. Thus α is separable over F if and only if $\mathrm{irr}(\alpha, F)$ has all zeros of multiplicity 1. This tells us at once that an irreducible polynomial $f(x) \in F[x]$ is separable over F if and only if $f(x)$ has all zeros of multiplicity 1.

Theorem 5.4.6. *Let K be a finite extension of E and E a finite extension of F. Then K is separable over F if and only if K is separable over E and E is separable over F.*

Proof. Note that

$$[K : F] = [K : E][E : F], \quad \{K : F\} = \{K : E\}\{E : F\},$$
$$\{K : F\} \mid [K : E], \quad \{E : F\} \mid [E : F].$$

Then K is separable over F, if and only if $[K : F] = \{K : F\}$, if and only if $[K : E][E : F] = \{K : E\}\{E : F\}$, if and only if $[K : E] = \{K : E\}$ and $[E : F] = \{E : F\}$ (Theorem 5.4.4). $\qquad\square$

Theorem 5.4.6 can be extended in the obvious way, by induction, to any finite tower of finite extensions. The top field is a separable extension of the bottom one if and only if each field is a separable extension of the one immediately under it.

Corollary 5.4.7. *Let E be a finite extension of F. Then E is separable over F if and only if each $\alpha \in E$ is separable over F.*

Proof. ($\Rightarrow$). Suppose that E is separable over F, and let $\alpha \in E$. Then

$$F \leq F(\alpha) \leq E,$$

and Theorem 5.4.6 shows that $F(\alpha)$ is separable over F, i.e., α in E is separable over F.

($\Leftarrow$). Suppose that every $\alpha \in E$ is separable over F. Since E is a finite extension of F, there exist $\alpha_1, \cdots, \alpha_n$ such that

$$F < F(\alpha_1) < F(\alpha_1, \alpha_2) < \cdots < E = F(\alpha_1, \cdots, \alpha_n).$$

Now since α_k is separable over F, α_k is separable over $F(\alpha_1, \cdots, \alpha_{k-1})$, because $q(x) = \mathrm{irr}(\alpha_k, F(\alpha_1, \cdots, \alpha_{k-1}))$ divides $\mathrm{irr}(\alpha_k, F)$, so that α_k is a zero of $q(x)$ of multiplicity 1. Thus $F(\alpha_1, \cdots, \alpha_k)$ is separable over $F(\alpha_1, \cdots, \alpha_{k-1})$, so E is separable over F by Theorem 5.4.6. So E is separable over F. $\qquad\square$

We next prove that α can fail to be separable over F only if F is an infinite field of characteristic $p \neq 0$.

Theorem 5.4.8. *Let $f(x) \in F[x]$ be an irreducible polynomial and let E be a splitting field for $f(x)$.*

(1). If F has characteristic zero, then $f(x)$ does not have multiple roots in E.

(2). If F has characteristic p and $f(x)$ has multiple roots in E, then $f(x) = g(x^p)$ for some $g(x) \in F[x]$.

Proof. (1) Since $f(x)$ is irreducible, $\deg f(x) \geq 1$. So since F has characteristic zero, $f'(x) \neq 0$. Because $f(x)$ is irreducible, we must then have that $\gcd(f(x), f'(x)) = 1$. So $f(x)$ cannot have multiple roots.

(2) By Theorem 4.2.10 we know that $f(x)$ has multiple roots in E if and only if $\gcd(f(x), f'(x)) \neq 1$, if and only if $\gcd(f(x), f'(x)) \sim f(x)$ since $f(x)$ is irreducible, if and only if $f'(x) = 0$, if and only if $f(x) = g(x^p)$ for some $g(x) \in F[x]$. $\qquad\square$

Definition 5.4.9. *A field F is* **perfect** *if every finite extension of F is separable.*

Theorem 5.4.10. *Every field of characteristic zero is perfect.*

Proof. Let E be a finite extension of a field F of characteristic zero, and let $\alpha \in E$. From Theorem 5.4.8 then $f(x) = \mathrm{irr}(\alpha, F)$ does not have multiple roots in $\overline{F}$. Therefore, α is separable over F for all $\alpha \in E$. By Corollary 5.4.7, this means that E is a separable extension of F. $\qquad\square$

We will find the answer for a field F of characteristic $p > 0$ to be perfect.

Theorem 5.4.11. *Let F be a field of characteristic p such that every element of F is a pth power, and $f(x) \in F[x] \setminus F$. Then $f(x^p)$ is not irreducible in $F[x]$.*

Proof. We know that $F = \{a^p : a \in F\}$.

Let $f(x) = a_n x^n + a_{n-1} x^{n-1} + \cdots + a_1 x + a_0$ where $a_i \in F$. Then $a_i = b_i^p$ for some $b_i \in F$. We see that

$$f(x^p) = b_n^p x^{pn} + b_{n-1}^p x^{p(n-1)} + \cdots + b_1^p x^p + b_0^p$$
$$= (b_n x^n + b_{n-1} x^{n-1} + \cdots + b_1 x + b_0)^p.$$

Thus $f(x^p)$ is not irreducible in $F[x]$. $\qquad\square$

Theorem 5.4.12. *A field F of characteristic $p > 0$ is perfect if and only if every element of F is a pth power.*

Proof. ($\Leftarrow$). Suppose that $F = \{a^p : a \in F\}$. Let E be a finite extension of F, and let $\alpha \in E$. Since $f(x) = \mathrm{irr}(\alpha, F)$ is irreducible, from Theorems 5.4.8 and 5.4.11 we know that it does not have multiple roots. Therefore, α is separable over F for all $\alpha \in E$. By Corollary 5.4.7, this means that E is a separable extension of F. Consequently, F is perfect.

($\Rightarrow$). Now suppose that $F \neq \{a^p : a \in F\}$. Take $a \in F$ which is not a prime of any element in F. Let E be the splitting field of $x^p - a$ over F. Take a root $\alpha \in E$ of $x^p - a$. We know that $a = \alpha^p$ and $x^p - a = (x - \alpha)^p$. Since $\alpha \notin F$, $\deg(\alpha, F) > 1$ and $\mathrm{irr}(\alpha, F) | (x - \alpha)^p$. Thus $\mathrm{irr}(\alpha, F)$ must have multiple root α, i.e., α is not separable over F. Consequently, F is not perfect. $\square$

As a consequence we have the following result.

Theorem 5.4.13. *Every finite field F is perfect.*

Proof. This follows from the fact that $F = \{a^p : a \in F\}$ (Theorem 5.1.12) and Theorem 5.4.12. $\square$

We have completed our aim: for finite extensions E of such perfect fields F, $[E : F] = \{E : F\}$ if and only if E is a separable extension.

The following theorem is very useful.

Theorem 5.4.14. *If E is a finite separable extension of a field F, then $E = F(\alpha)$ for some $\alpha \in E$.*

Proof. If F is a finite field, then E is also finite. Let α be a generator for the cyclic group E^* of nonzero elements of E under multiplication. Clearly, $E = F(\alpha)$, so α is a primitive element in this case.

We now assume that F is infinite, and prove our theorem in the case that $E = F(\beta, \gamma)$. The induction argument from this to the general case is straightforward. Let $f(x) = \mathrm{irr}(\beta, F)$ have distinct zeros $\beta = \beta_1, \cdots, \beta_n$, and let $\mathrm{irr}(\gamma, F)$ have distinct zeros $\gamma = \gamma_1, \cdots, \gamma_m$ in $\overline{F}$, where all zeros are of multiplicity 1, since E is a separable extension of F. Since F is infinite, we can find $a \in F$ such that

$$a \neq (\beta_i - \beta)/(\gamma - \gamma_j)$$

for all i and j, with $j \neq 1$. That is, $a(\gamma - \gamma_j) \neq \beta_i - \beta$. Letting $\alpha = \beta + a\gamma$, we have

$$\alpha = \beta + a\gamma \neq \beta_i + a\gamma_j,$$

so $\alpha - a\gamma_j \neq \beta_i$ for all i and all $j \neq 1$. Consider

$$h(x) = f(\alpha - ax) \in (F(\alpha))[x].$$

Now $h(\gamma) = f(\beta) = 0$. However, $h(\gamma_j) \neq 0$ for $j \neq 1$ by construction, since the β_i were the only zeros of $f(x)$. Hence $h(x)$

and $g(x) = \mathrm{irr}(\gamma, F)$ have a common factor in $(F(\alpha))[x]$, namely $\mathrm{irr}(\gamma, F(\alpha))$, which must be linear, since γ is the only common zero of $g(x)$ and $h(x)$. Thus $\gamma \in F(\alpha)$, and therefore $\beta = \alpha - a\gamma$ is in $F(\alpha)$. Hence $F(\beta, \gamma) = F(\alpha)$. $\qquad\square$

Corollary 5.4.15. *Any finite extension of a field of characteristic zero is a simple extension.*

Proof. This corollary follows at once from Theorems 5.4.10 and 5.4.14. $\qquad\square$

We see that the only possible "bad case" where a finite extension may not be simple is a finite extension of an infinite field of characteristic $p \neq 0$.

Example 5.4.3. *Let p be a prime, $L = F_p(x_1, x_2)$, the fraction field of the polynomial ring $F_p[x_1, x_2]$. Let $K = F_p(x_1^p, x_2^p)$. Show that L is not a simple extension over K.*

Proof. If $\alpha \in L$, there are $f(x_1, x_2), g(x_1, x_2) \in F_p[x_1, x_2]$ such that $\alpha = \frac{f(x_1, x_2)}{g(x_1, x_2)}$. Then

$$\alpha^p = \frac{f(x_1, x_2)^p}{g(x_1, x_2)^p} = \frac{f(x_1^p, x_2^p)}{g(x_1^p, x_2^p)} \in K$$

since $a^p = a$ for any $a \in F_p$. So α is a root of $x^p - \alpha^p \in K[x]$. Thus the irreducible polynomial $\mathrm{irr}(\alpha, K)$ has degree at most p. So $[K(\alpha) : K] = \deg(\mathrm{irr}(\alpha, K)) \leq p$. On the other hand, we have $[L : K] = p^2$, since $\{x_1^k x_2^j : 0 \leq k, j < p\}$ is a basis. For any $\alpha \in L$, we have $K(\alpha) \neq L$. So $K \leq L$ is not a simple extension. This then implies $K \leq L$ is not separable. $\qquad\square$

It is interesting to know the following useful result which is named after Jacob Lüroth (1844–1910), who proved it in 1876. We omit the proof while a detailed proof can be found in [M, Theorem 8.19].

Theorem 5.4.16 (Lüroth's Theorem). *Let $K \leq E \leq K(x)$ be fields where x is an indeterminate. Then there is $\alpha \in K(x)$ such that $E = K(\alpha)$.*

5.5. Exercises.

(1) Find the centralizer of complex conjugation in $\mathrm{Aut}(\mathbb{C}/\mathbb{Q})$.
(2) (a). Show that $x^4 + 4$ is reducible over $\mathbb{Q}$.
    (b). Find the splitting field over $\mathbb{Q}$ for the polynomial $x^4 + 4$.

(c). Find the Galois group over $\mathbb{Q}$ of the polynomial $x^4 + 4$.

(3) Let F be a field generated over the field K by u and v of relatively prime degrees m and n, respectively, over K. Prove that $[F : K] = mn$.

(4) Let $F \supseteq K$ be an extension field, with $u \in F$. If $[K(u) : K]$ is an odd number, show that $K(u^2) = K(u)$.

(5) Find the degree $[F : \mathbb{Q}]$, where F is the splitting field of the polynomial $x^3 - 11$ over the field $\mathbb{Q}$ of rational numbers.

(6) Let $f(x) \in \mathbb{Q}[x]$ be irreducible over $\mathbb{Q}$, and let F be the splitting field for $f(x)$ over $\mathbb{Q}$. If $[F : \mathbb{Q}]$ is odd, prove that all of the roots of $f(x)$ are real.

(7) A field F is called formally real if -1 is not expressible as a sum of squares in F. Let $f(x) \in F[x]$ be an irreducible polynomial of odd degree, and let α be a root of $f(x)$. Prove that if F is formally real, then $F[\alpha]$ is also formally real.

(8) Let K_1, K_2 be finite extensions of a field F contained in the field K, and assume both are splitting field over F. Prove that K_1K_2 and $K_1 \cap K_2$ are splitting field over F.

(9) Calculate the splitting field E of $f(x) = x^3 + x + 1$ over $\mathbb{Z}_2$, and find $|E|$.

(10) Find the degree of the splitting field over $\mathbb{Z}_2$ for the polynomial $(x^3 + x + 1)(x^2 + x + 1)$.

(11) Determine the splitting field over $\mathbb{Q}$ for $x^4 + x^2 + 1$.

(12) Determine the splitting field over $\mathbb{Q}$ for $x^4 + 2$.

(13) Find the splitting field for $x^3 + x + 1$ over $\mathbb{Z}_2$.

(14) Let p be a prime number. Find the splitting fields for $x^p - 1$ over $\mathbb{Q}$ and over $\mathbb{R}$.

(15) Find the splitting field of $x^6 - 1 \in \mathbb{Z}_5[x]$ over $\mathbb{Z}_5$.

(16) Find the splitting field of $x^5 - 1 \in \mathbb{Z}_2[x]$ over $\mathbb{Z}_2$.

(17) Find the degree of the splitting field over $\mathbb{Z}_2$ for the polynomial $(x^3 + x + 1)(x^2 + x + 1)$.

(18) Let F be the splitting field in $\mathbb{Q}$ of $x^4 + 1$.
  (a). Show that $[F : \mathbb{Q}] = 4$.
  (b). Find automorphisms of F that have fixed fields $\mathbb{Q}(\sqrt{2})$, $\mathbb{Q}(i)$, and $\mathbb{Q}(\sqrt{2}i)$, respectively.

(19) Let $\overline{F}$ be an algebraic closure of F, and let
$$f(x) = x^n + a_{n-1}x^{n-1} + \cdots + a_1x + a_0 \in \overline{F}[x].$$

If $(f(x))^m \in F[x]$ and $m \cdot 1 \neq 0$ in F, show that $f(x) \in F[x]$, that is, all $a_i \in F$.

(20) Let F be a field, $u = f(x)/g(x) \in F[x] \setminus F$ where $\gcd(f(x), g(x)) = 1$. Show u is transcendental over F, x is algebraic over $F[u]$ and $[F(x) : F(u)] = \max(\deg(f(x)), \deg(f(x)))$.

(21) Let I be a maximal ideal in $F[x_1, \cdots, x_n]$, where F is a field. Show that the field $F[x_1, \cdots, x_n]/I$ is a finite extension of F.

(22) Show that $\sigma \in \mathrm{Gal}(F(x)/F)$ if and only if

$$\sigma(x) = \frac{ax + b}{cx + d}, a, b, c, d \in F$$

with $ad - bc \neq 0$.

(Remark: The group $\mathrm{Gal}(F(x_1, x_2, \cdots, x_n)/F)$ is quite complicated for $n = 2$ and is unknown for $n \geq 3$.)

6. GALOIS THEORY

In 1830, Evariste Galois (1811–1832), at the age of 18, submitted to the Paris Academy of Sciences a memoir on his theory of solvability by radicals; Galois' paper was ultimately rejected in 1831 as being too sketchy and for giving a condition in terms of the roots of the equation instead of its coefficients. Galois then died in a duel in 1832, and his paper, "Mémoire sur les conditions de résolubilité des équations par radicaux", remained unpublished until 1846 when it was published by Joseph Liouville (1809–1882) accompanied by some of his own explanations. Prior to this publication, Liouville announced Galois' result to the Academy in a speech he gave on July 4, 1843. Galois's characterization "dramatically supersedes the work of Abel and Ruffini." (Cited from Wikipedia.)

In this last chapter we will prove the elegant Galois Theorem that explains the relationship between subfields of a field and its automorphism group. As an application we provide the necessary and sufficient conditions for polynomial equations over a field of characteristic 0 in one variable to be solvable in terms of nested roots.

6.1. Galois Theorem.

The Galois theory gives a beautiful interplay of group theory and field theory. We shall start by recalling the main results we have developed and should keep well in mind. We always assume that $\overline{F}$ is the algebraic closure of the field F.

1. Let $F \leq E \leq \overline{F}, \alpha \in E$, and let β be a conjugate of α over F, that is, $\operatorname{irr}(\alpha, F) = \operatorname{irr}(\beta, F)$. Then we have the conjugation isomorphism $\psi_{\alpha,\beta} : F(\alpha) \to F(\beta)$ that leaves F fixed and maps α onto β.

2. If $F \leq E \leq \overline{F}$ and $\alpha \in E$, then any $\sigma \in \operatorname{Gal}(\overline{F}/F)$ maps α onto some conjugate of α.

3. If $F \leq E$, we have the Galois group $\operatorname{Gal}(E/F)$. For any $S \subseteq \operatorname{Gal}(E/F)$, $E_S \leq E$. In particular, $E_{\operatorname{Gal}(E/F)} \leq E$.

4. A field E, $F \leq E \leq \overline{F}$, is a splitting field over F if and only if $S(E/F) \leq \operatorname{Gal}(E/F)$. If E is a finite extension and a splitting field over F, then $|\operatorname{Gal}(E/F)| = \{E : F\}$.

5. If E is a finite extension of F, then $\{E : F\}|[E : F]$. If E is also separable over F, then $\{E : F\} = [E : F]$. Also, E is separable over F if and only if irr(α, F) has all zeros of multiplicity 1 for every $\alpha \in E$.

6. If E is a finite extension of F and is a separable splitting field over F, then $|\mathrm{Gal}(E/F)| = \{E : F\} = [E : F]$.

Definition 6.1.1. *A finite extension K of F is a **finite normal extension of** F if K is a separable splitting field over F.*

A finite normal extension of a field F is also called a **Galois extension** of F.

Theorem 6.1.2. *Let K be a finite normal extension of F, and let E be an extension of F such that $F \leq E \leq K \leq \overline{F}$.*

(a). K is a finite normal extension of E, and

$$\mathrm{Gal}(K/E) = \{\sigma \in \mathrm{Gal}(K/F) : \sigma(a) = a \; \forall a \in E\}.$$

(b). $\sigma, \tau \in \mathrm{Gal}(K/F)$ induce the same isomorphism of E onto a subfield of $\overline{F}$ if and only if they are in the same left coset of $\mathrm{Gal}(K/E)$ in $\mathrm{Gal}(K/F)$.

Proof. (a). If K is the splitting field of a set $P = \{f_i(x)|i \in I\}$ of polynomials in $F[x]$, then K is the splitting field over E of $P \subset E[x]$. Since K is separable over F then K is separable over E. Thus K is a normal extension of E. The first statement follows.

For any $\sigma \in \mathrm{Gal}(K/E)$ we have $\sigma(a) = a$ for any $a \in F$. Thus $\mathrm{Gal}(K/E) \subseteq \mathrm{Gal}(K/F)$. Since $\mathrm{Gal}(K/E)$ is a group under function composition also, we see that $\mathrm{Gal}(K/E) \leq \mathrm{Gal}(K/F)$.

(b). Let $\sigma, \tau \in \mathrm{Gal}(K/F)$. Then σ and τ are in the same left coset of $\mathrm{Gal}(K/E)$ if and only if $\mu = \tau^{-1}\sigma \in \mathrm{Gal}(K/E)$, if and only if

$$\sigma(\alpha) = (\tau\mu)(\alpha) = \tau(\mu(\alpha)) = \tau(\alpha),$$

since $\mu(\alpha) = \alpha$ for $\alpha \in E$. □

If $F \leq E \leq K$ is a chain of field extensions such that $F \leq E$ is normal and $E \leq K$ is normal, it is false to conclude that $F \leq K$ is normal. For example $\mathbb{Q} \leq \mathbb{Q}[\sqrt{2}] \leq \mathbb{Q}[\sqrt{\sqrt{2} - 1}]$.

Example 6.1.1. *Let $E = \mathbb{Q}(\sqrt{2}, \sqrt{3})$. Find all subgroups of $\mathrm{Gal}(E/\mathbb{Q})$ and their corresponding fixed subfields of E.*

Solution. Now E is a finite normal extension of $\mathbb{Q}$, and a previous example showed that $|\text{Gal}(E/\mathbb{Q})| = 4$. We recall them by giving their values on the basis $\{1, \sqrt{2}, \sqrt{3}, \sqrt{6}\}$ for E over $\mathbb{Q}$.

ι : The identity map,

$\sigma_1 : \sqrt{2} \to -\sqrt{2}, \sqrt{6} \to -\sqrt{6}$, and leaves the other basis elements fixed,

$\sigma_2 : \sqrt{3} \to -\sqrt{3}, \sqrt{6} \to -\sqrt{6}$, and leaves the other basis elements fixed,

$\sigma_3 : \sqrt{2} \to -\sqrt{2}, \sqrt{3} \to -\sqrt{3}$, and leaves the other basis elements fixed.

We saw that $\{\iota, \sigma_1, \sigma_2, \sigma_3\}$ is isomorphic to $\mathbb{Z}_2 \times \mathbb{Z}_2$. The complete list of subgroups, with each subgroup paired off with the corresponding intermediate field that it leaves fixed, is as follows:

$$\{\iota, \sigma_1, \sigma_2, \sigma_3\} \iff \mathbb{Q},$$
$$\{\iota, \sigma_1\} \iff \mathbb{Q}(\sqrt{3}),$$
$$\{\iota, \sigma_2\} \iff \mathbb{Q}(\sqrt{2}),$$
$$\{\iota, \sigma_3\} \iff \mathbb{Q}(\sqrt{6}),$$
$$\{\iota\} \iff \mathbb{Q}(\sqrt{2}, \sqrt{3}).$$

The diagram of subgroups and the diagram of subfields are as follows.

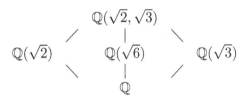

All subgroups of the abelian group $\{\iota, \sigma_1, \sigma_2, \sigma_3\}$ are normal subgroups, and all the intermediate fields are normal extensions of $\mathbb{Q}$. $\square$

When K is a finite normal extension of F, and G a group, we define the following sets

$$[\![K, F]\!] = \{E : F \leq E \leq K\}, \quad [\![G]\!] = \{H : H \leq G\}.$$

Them we have the following maps:

$$\begin{aligned} \lambda : &[\![K, F]\!] \to [\![\mathrm{Gal}(K/E)]\!], \quad E \mapsto \mathrm{Gal}(K/E), \forall E \in [\![K, F]\!]; \\ \mu : &[\![\mathrm{Gal}(K/E)]\!] \to [\![K, F]\!], \quad H \mapsto E_H, \forall H \in [\![\mathrm{Gal}(K/E)]\!]. \end{aligned} \quad (6.1)$$

We will prove our main theorem in a few steps.

Theorem 6.1.3. *Let K be a finite normal extension of a field F, let $E \in [\![K, F]\!]$.*

(1). $\lambda : [\![K, F]\!] \to [\![\mathrm{Gal}(K/E)]\!]$ *is a one-to-one map, and $E = K_{\lambda(E)}$.*

(2). $[K : E] = |\mathrm{Gal}(K/E)|$ *and $[E : F] = (\mathrm{Gal}(K/F) : \mathrm{Gal}(K/E))$, the number of left cosets of $\mathrm{Gal}(K/E)$ in $\mathrm{Gal}(K/F)$.*

Proof. (1). By definition we know that $E \leq K_{\mathrm{Gal}(K/E)}$. Next we prove that $K_{\mathrm{Gal}(K/E)} \leq E$.

Let $\alpha \in K \setminus E$. Since K is a normal extension of E, by using a conjugation isomorphism and the Isomorphism Extension Theorem, we can find an automorphism of K leaving E fixed and mapping α onto a different zero of $\mathrm{irr}(\alpha, E)$. So $\alpha \notin K_{\mathrm{Gal}(K/E)}$. This implies that $K_{\mathrm{Gal}(K/E)} \leq E$, so $E = K_{\mathrm{Gal}(K/E)}$.

For any $E_1, E_2 \in [\![K, F]\!]$, if $\lambda(E_1) = \lambda(E_2)$, then

$$E_1 = K_{\lambda(E_1)} = K_{\lambda(E_2)} = E_2.$$

So λ is one to one.

(2). Since K is a finite normal extension over F and E, from Corollary 5.3.5 we see that $[K : E] = \{K : E\} = |S(K/E)| = |\mathrm{Gal}(K/E)|$. Again from Corollary 5.3.5 we obtain that

$$[E : F] = [K : F]/[K : E] = \{K : F\}/\{K : E\} = |S(K/F)|/|S(K/E)|$$
$$= |\mathrm{Gal}(K/F)|/|\mathrm{Gal}(K/E)| = (\mathrm{Gal}(K/F) : \mathrm{Gal}(K/E)).$$

$\square$

Theorem 6.1.4. *Let K be a finite normal extension of a field F. For any $G \leq \mathrm{Gal}(K/F)$, we have $\mathrm{Gal}(K/K_G) = G$.*

Proof. Let $G \leq \mathrm{Gal}(K/F)$. Then $K_G = \{a \in K : \sigma(a) = a, \forall \sigma \in G\}$ and

$$G \leq \mathrm{Gal}(K/K_G) \leq \mathrm{Gal}(K/F).$$

We need to show that it is impossible to have G a proper subgroup of $\mathrm{Gal}(K/K_G)$. We shall suppose that $G < \mathrm{Gal}(K/K_G)$ and shall derive a contradiction. As a finite separable extension, $K = K_G(\alpha)$ for some $\alpha \in K$. Let

$$n = [K : K_G] = \{K : K_G\} = |\mathrm{Gal}(K/K_G)|.$$

Then $G < \mathrm{Gal}(K/K_G)$ implies that $r = |G| < |\mathrm{Gal}(K/K_G)| = n$. Let $G = \{\sigma_1, \cdots, \sigma_r\}$, and consider the polynomial

$$f(x) = \prod_{i=1}^{r}(x - \sigma_i(\alpha)).$$

Then $f(x)$ is of degree $r < n$. Now the coefficients of each power of x in $f(x)$ are symmetric expressions in the $\sigma_i(\alpha)$. For example, the coefficient of x^{r-1} is

$$-\sigma_1(\alpha) - \sigma_2(\alpha) - \cdots - \sigma_r(\alpha).$$

Thus these coefficients are invariant under each isomorphism $\sigma_i \in G$, since $G = \{\sigma\sigma_1, \cdots, \sigma\sigma_r\}$ for any $\sigma \in G$. Hence $f(x) \in K_G[x]$. Since some σ_i is ι, we see that some $\sigma_i(\alpha)$ is α, so $f(\alpha) = 0$. So $\mathrm{irr}(\alpha, K_G)|f(x)$. Therefore, we would have

$$\deg(\alpha, K_G) \leq r < n = [K : K_G] = [K_G(\alpha) : K_G] = \deg(\alpha, K_G).$$

This is impossible. Thus we have proved our result.   □

Theorem 6.1.5. *Let K be a finite normal extension of a field F, let $E \in [\![K, F]\!]$.*

   (1). E is a normal extension of F if and only if $\mathrm{Gal}(K/E) \trianglelefteq \mathrm{Gal}(K/F)$.

   (2). If E is a normal extension of F, then $\mathrm{Gal}(E/F) \simeq \mathrm{Gal}(K/F)/ \mathrm{Gal}(K/E)$.

Proof. (1). Since $E \in [\![K, F]\!]$, it is separable over F. Thus E is normal over F if and only if E is a splitting field over F. By the Isomorphism Extension Theorem, every isomorphism of E onto a subfield of $\overline{F}$ leaving F fixed can be extended to an automorphism of K, since K is normal over F. Thus the automorphisms of $\mathrm{Gal}(K/F)$ induce all possible isomorphisms of E onto a subfield of $\overline{F}$ over F. We know that E is a splitting field over F, (hence is normal over F), if and only if

$$\sigma(\alpha) \in E, \forall \sigma \in \mathrm{Gal}(K/F), \alpha \in E,$$

if and only if

$$\tau(\sigma(\alpha)) = \sigma(\alpha), \forall \sigma \in \text{Gal}(K/F), \alpha \in E, \tau \in \text{Gal}(K/E),$$

if and only if

$$(\sigma^{-1}\tau\sigma)(\alpha) = \alpha, \forall \alpha \in E, \sigma \in \text{Gal}(K/F), \text{ and } \tau \in \text{Gal}(K/E),$$

if and only if

$$(\sigma^{-1}\tau\sigma) \in \text{Gal}(K/E), \forall \sigma \in \text{Gal}(K/F), \text{ and } \tau \in \text{Gal}(K/E),$$

if and only if $\text{Gal}(K/E) \trianglelefteq \text{Gal}(K/F)$.

(2). For $\sigma \in \text{Gal}(K/F)$, let σ_E be the automorphism of E induced by σ (we are assuming that E is a normal extension of F). Thus $\sigma_E \in \text{Gal}(E/F)$. The map

$$\phi : \text{Gal}(K/F) \to \text{Gal}(E/F), \sigma \mapsto \sigma_E$$

is a homomorphism. By the Isomorphism Extension Theorem, every $\mu \in \text{Gal}(E/F)$ can be extended to some automorphism of K; that is, it is $\mu = \tau_E$ for some $\tau \in \text{Gal}(K/F)$. Thus ϕ is onto $\text{Gal}(E/F)$. The kernel of ϕ is $\text{Gal}(K/E)$. Therefore, by the First Isomorphism Theorem, $\text{Gal}(E/F) \simeq \text{Gal}(K/F)/\text{Gal}(K/E)$. $\square$

We have proved the Galois Theorem.

Theorem 6.1.6 (Galois Theorem). *Let K be a finite normal extension of a field F, let $E \in [\![K, F]\!]$.*

(1). The maps λ and μ defined in (6.1) are bijections and $\lambda = \mu^{-1}$.

(2). $[K : E] = |\text{Gal}(K/E)|$ and $[E : F] = (\text{Gal}(K/F) : \text{Gal}(K/E))$.

(3). E is a normal extension of F if and only if $\text{Gal}(K/E) \trianglelefteq \text{Gal}(K/F)$.

(4). If E is a normal extension of F, then $\text{Gal}(E/F) \simeq \text{Gal}(K/F)/\text{Gal}(K/E)$.

The Main Theorem of Galois Theory is a strong tool in the study of zeros of polynomials. If $f(x) \in F[x]$ is such that every irreducible factor of $f(x)$ is separable over F, then the splitting field K of $f(x)$ over F is a normal extension of F. The Galois group $\text{Gal}(K/F)$ is the **(Galois) group of the polynomial** $f(x)$ **over** F which is also denoted by $\text{Gal}(f(x)/F)$, even for any $f(x) \in F[x]$.

The following lemma is obvious.

Lemma 6.1.7. *Let K be a splitting field for $f(x) \in F[x]$ over F, let $\mathcal{R}_{f(x)} = \{\alpha_1, \cdots, \alpha_n\}$. Then $K = F(\alpha_1, \cdots, \alpha_n)$, and any $\sigma \in$ $\mathrm{Gal}(K/F)$ permutes the roots $\alpha_1, \cdots, \alpha_n$ so gives an injective group homomorphism $\mathrm{Gal}(K/F) \to S_n$, i.e., $\mathrm{Gal}(K/F) \le S_n$.*

We can determine the Galois groups of finite fields.

Theorem 6.1.8. *Let K be a finite extension of degree n of a finite field F of p^r elements. Then $\mathrm{Gal}(K/F)$ is cyclic of order n, and is generated by σ_{p^r}, where for $\alpha \in K, \sigma_{p^r}(\alpha) = \alpha^{p^r}$.*

Proof. From Theorem 5.4.13 (any finite field is perfect) we have seen that K is a separable extension of F. Since $|F| = p^r$ and $[K : F] = n$, so $|K| = p^{rn}$. Then we have seen that K is the splitting field of $x^{p^{rn}} - x$ over F. Hence K is a normal extension of F.

From Theorem 4.4.5 we know that $K_{\sigma_{p^r}} = \{a \in K : a^{p^r} = a\} = F$, and $(\sigma_{p^r})^k = \iota$ implies that $n|k$. So $|\langle \sigma_{p^r} \rangle| = n$. Since $|\mathrm{Gal}(K/F)| = [K : F] = n$, then $\mathrm{Gal}(K/F) = \langle \sigma_{p^r} \rangle$ is cyclic and generated by σ_{p^r}. $\square$

We use this theorem to give another illustration of the Main Theorem of Galois Theory.

Example 6.1.2. *Let $F = \mathbb{Z}_p$, and let $E = F_{p^{12}}$, so $[E : F] = 12$. Then $\mathrm{Gal}(E/F)$ is isomorphic to the cyclic group $\langle \mathbb{Z}_{12}, + \rangle$. All the subgroups of $\mathrm{Gal}(E/F) = \langle \sigma_p \rangle$ are the following:*

$$\langle \sigma_p \rangle, \langle \sigma_p^2 \rangle, \langle \sigma_p^3 \rangle, \langle \sigma_p^4 \rangle, \langle \sigma_p^6 \rangle, \langle \sigma_p^{12} \rangle.$$

For example,

$$\langle \sigma_p^4 \rangle = \{\iota, \sigma_p^4, \sigma_p^8\}.$$

6.2. Examples and an application.

In this section we will give some examples to compute the Galois groups of some polynomials and use the Galois Theorem to prove the fundamental theorem of algebra.

Example 6.2.1. *Let E be the splitting field of $x^3 - 2$ over $\mathbb{Q}$. Find all subgroups of $\mathrm{Gal}(E/\mathbb{Q})$ and their corresponding fixed subfields.*

Solution. From Example 5.3.2 we know that $E = \mathbb{Q}(\sqrt[3]{2}, \sqrt[3]{2}\omega, \sqrt[3]{2}\omega^2) = \mathbb{Q}(\sqrt[3]{2}, \omega)$, where $\omega = e^{\frac{2\pi i}{3}}$, which is the splitting field of $x^3 - 2$. So E is finite normal over $\mathbb{Q}$. Any element of $G = \mathrm{Gal}(E/\mathbb{Q})$ permutes $\sqrt[3]{2}, \sqrt[3]{2}\omega, \sqrt[3]{2}\omega^2$. From Lemma 6.1.7 we know that $G \subseteq S_3$.

Since $|G| = [E : \mathbb{Q}] = 6$, we must have $G = S_3$. All the subgroups of G are as follows.

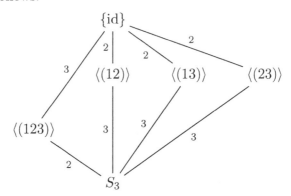

It is easy to compute all the corresponding subfields of G as follows.

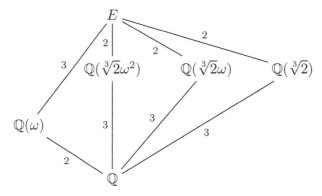

$\square$

Example 6.2.2. *Let E be the splitting field of $x^4 - 2$ over $\mathbb{Q}$. Find all subgroups of $\mathrm{Gal}(K/\mathbb{Q})$ and their corresponding fixed subfields.*

Solution. Now $x^4 - 2$ is irreducible over $\mathbb{Q}$ by Schönemann-Eisenstein Criterion. Let $\alpha = \sqrt[4]{2}$. Then the four zeros of $x^4 - 2$ in $\mathbb{C}$ are $\alpha, -\alpha, i\alpha$, and $-i\alpha$. We see that $(i\alpha)/\alpha = i \in K$. Since α is a real number, $\mathbb{Q}(\alpha) < \mathbb{R}$, so $\mathbb{Q}(\alpha) \neq K$. However, since $\mathbb{Q}(\alpha, i)$ contains all zeros of $x^4 - 2$, we see that $\mathbb{Q}(\alpha, i) = K$. Letting $E = \mathbb{Q}(\alpha)$.

Now $\{1, \alpha, \alpha^2, \alpha^3\}$ is a basis for E over $\mathbb{Q}$, and $\{1, i\}$ is a basis for K over E. Thus $\{1, \alpha, \alpha^2, \alpha^3, i, i\alpha, i\alpha^2, i\alpha^3\}$ is a basis for K over $\mathbb{Q}$. From Lemma 6.1.7 we know that $\mathrm{Gal}(K/\mathbb{Q}) \subseteq S_4$. Since $[K : \mathbb{Q}] = 8$, we must have $|\mathrm{Gal}(K/\mathbb{Q})| = 8$. We will find eight automorphisms $\sigma \in \mathrm{Gal}(K/\mathbb{Q})$.

Since $K = \mathbb{Q}(\alpha, i)$, each $\text{Gal}(K/\mathbb{Q})$ is completely determined by $\sigma(i)$ and $\sigma(\alpha)$. But $\sigma(\alpha) \in \{1, \alpha, \alpha^2, \alpha^3\}$. Likewise, $\sigma(i) = \pm i$, zeros of $\text{irr}(i, \mathbb{Q}) = x^2 + 1$. Thus the four possibilities for $\sigma(\alpha)$, combined with the two possibilities for $\sigma(i)$, must give all eight automorphisms.

	ρ_0	ρ_1	ρ_2	ρ_3	μ_1	δ_1	μ_2	δ_2
$\alpha \mapsto$	α	$i\alpha$	$-\alpha$	$-i\alpha$	α	$i\alpha$	$-\alpha$	$-i\alpha$
$i \mapsto$	i	i	i	i	$-i$	$-i$	$-i$	$-i$

For example, $\rho_3(\alpha) = -i\alpha$ and $\rho_3(i) = i$, while ρ_0 is the identity automorphism.

Now

$$(\mu_1\rho_1)(\alpha) = \mu_1(\rho_1(\alpha)) = \mu_1(i\alpha) = \mu_1(i)\mu_1(\alpha) = -i\alpha,$$

and, similarly,

$$(\mu_1\rho_1)(i) = -i,$$

so $\mu_1\rho_1 = \delta_2$. A similar computation shows that

$$(\rho_1\mu_1)(\alpha) = i\alpha \text{ and } (\rho_1\mu_1)(i) = -i.$$

Thus $\rho_1\mu_1 = \delta_1$, so $\rho_1\mu_1 \neq \mu_1\rho_1$ and $\text{Gal}(K/\mathbb{Q})$ is not abelian.

There are two noisomorphic nonabelian groups of order 8. Since ρ_1 is of order 4, μ_1 is of order 2, $\{\rho_1, \mu_1\}$ generates $\text{Gal}(K/\mathbb{Q})$, and $\mu_1\rho_1\mu_1 = \rho_1^{-1}$. Thus $\text{Gal}(K/\mathbb{Q})$ is isomorphic to the dihedral group D_4 (also called octic group). Actually all the subgroups of $\text{Gal}(K/\mathbb{Q})$ are

$$H_0 = \text{Gal}(K/\mathbb{Q}), \qquad H_1 = \{\rho_0, \rho_2, \mu_1, \mu_2\},$$
$$H_2 = \{\rho_0, \rho_1, \rho_2, \rho_3\}, \quad H_3 = \{\rho_0, \rho_2, \delta_1, \delta_2\},$$
$$H_4 = \{\rho_0, \mu_1\}, \qquad H_5 = \{\rho_0, \mu_2\},$$
$$H_6 = \{\rho_0, \rho_2\}, \qquad H_7 = \{\rho_0, \delta_1\},$$
$$H_8 = \{\rho_0, \delta_2\}, \qquad H_9 = \{\rho_0\}.$$

The determination of the fixed fields K_{H_i} sometimes requires a bit of ingenuity. Let's illustrate.

To find K_{H_2}, we merely have to find an extension of $\mathbb{Q}$ of degree 2 left fixed by $\{\rho_0, \rho_1, \rho_2, \rho_3\}$. Since all ρ_j leave i fixed, then $K_{H_2} = \mathbb{Q}(i)$.

To find K_{H_4}, we have to find an extension of $\mathbb{Q}$ of degree 4 left fixed by ρ_0 and μ_1. Since μ_1 leaves α fixed and α is a zero of $\text{irr}(\alpha, \mathbb{Q}) = x^4 - 2$, we see that $\mathbb{Q}(\alpha)$ is of degree 4 over $\mathbb{Q}$ and is left fixed by $\{\rho_0, \mu_1\}$. By Galois Theorem 6.1.6, it is the only such field. Here we are using strongly the one-to-one correspondence given by the Galois theory. Then $K_{H_4} = \mathbb{Q}(\alpha)$.

Let us find K_{H_7}. Since $H_7 = \{\rho_0, \delta_1\}$ is a group, for any $\beta \in K$ we see that $\rho_0(\beta) + \delta_1(\beta)$ is left fixed by ρ_0 and δ_1. Taking $\beta = \alpha$, we see that $\rho_0(\alpha) + \delta_1(\alpha) = \alpha + i\alpha$ is left fixed by H_7. We can check and see that ρ_0 and δ_1 are the only automorphisms leaving $\alpha + i\alpha$ fixed. Thus, by the one-to-one correspondence, we must have

$$\mathbb{Q}(\alpha + i\alpha) = \mathbb{Q}(\sqrt[4]{2} + i\sqrt[4]{2}) = K_{H_7}.$$

Next we find $\mathrm{irr}(\alpha + i\alpha, \mathbb{Q})$. If $\gamma = \alpha + i\alpha$, then for every conjugate of γ over $\mathbb{Q}$, there exists an automorphism of K mapping γ into that conjugate. Thus we need only compute the various different values $\sigma(\gamma)$ for $\sigma \in \mathrm{Gal}(K/\mathbb{Q})$ to find the other zeros of $\mathrm{irr}(\gamma, \mathbb{Q})$. Elements σ of $\mathrm{Gal}(K/\mathbb{Q})$ giving these different values can be found by taking a set of representatives of the left cosets of $\mathrm{Gal}(K/\mathbb{Q}(\gamma)) = \{\rho_0, \delta_1\}$ in $\mathrm{Gal}(K/\mathbb{Q})$. A set of representatives for these left cosets is $\{\rho_0, \rho_1, \rho_2, \rho_3\}$.

The conjugates of $\gamma = \alpha + i\alpha$ are thus $\alpha + i\alpha, i\alpha - \alpha, -\alpha - i\alpha$, and $-i\alpha + \alpha$. Hence

$$\mathrm{irr}(\gamma, \mathbb{Q}) = [(x - (\alpha + i\alpha))(x - (i\alpha - \alpha))] \cdot [(x - (-\alpha - i\alpha))(x - (-i\alpha + \alpha))]$$

$$= (x^2 - 2i\alpha x - 2\alpha^2)(x^2 + 2i\alpha x - 2\alpha^2) = x^4 + 4\alpha^4 = x^4 + 8.$$

You may try to use a simpler way to compute $\mathrm{irr}(\gamma, \mathbb{Q})$.

We leave all other subgroups as exercises for you to compute their fixed subfields. □

Our next example will give an extension of degree 4 for the splitting field of a quartic.

Example 6.2.3. Let K be the splitting field of $x^4 + 1$ over $\mathbb{Q}$. Find all subgroups of $\mathrm{Gal}(K/\mathbb{Q})$ and their corresponding fixed subfields.

Solution. Since $x^4 + 1 = (x^2 - i)(x^2 + i)$ is the irreducible factorization over the field $\mathbb{Q}[i]$, which is not a factorization in $\mathbb{Q}[x]$, so $x^4 + 1$ is irreducible over $\mathbb{Q}$. The zeros of $x^4 + 1$ are $(1 \pm i)/\sqrt{2}$ and $(-1 \pm i)/\sqrt{2}$. A computation shows that if

$$\alpha = \frac{1 + i}{\sqrt{2}},$$

then

$$\alpha^2 = i, \alpha^3 = \frac{-1 + i}{\sqrt{2}}, \alpha^5 = \frac{-1 - i}{\sqrt{2}}, \text{ and } \alpha^7 = \frac{1 - i}{\sqrt{2}}.$$

Thus $K = \mathbb{Q}(\alpha)$, and $[K : \mathbb{Q}] = 4$. Let us compute $\text{Gal}(K/\mathbb{Q})$ and give the group. Since there exist automorphisms of K mapping α onto each conjugate of α, and since an automorphism σ of $\mathbb{Q}(\alpha)$ is completely determined by $\sigma(\alpha)$, we see that the four elements of $\text{Gal}(K/\mathbb{Q})$ are defined by

	σ_1	σ_3	σ_5	σ_7
$\alpha \mapsto$	α	α^3	α^5	α^7

Since

$$(\sigma_j \sigma_k)(\alpha) = \sigma_j(\alpha^k) = (\alpha^j)^k = \alpha^{jk}$$

and $\alpha^8 = 1$, we see that $\text{Gal}(K/\mathbb{Q})$ is isomorphic to the group $\{1, 3, 5, 7\}$ under multiplication modulo 8. We know that $\sigma_j^2 = \sigma_1$, the identity, for all j. We can see that $\text{Gal}(K/\mathbb{Q})$ is the Klein 4-group which is isomorphic to $\mathbb{Z}_2 \times \mathbb{Z}_2$.

To find $K_{\{\sigma_1,\sigma_3\}}$, it is only necessary to find an element of K not in $\mathbb{Q}$ left fixed by $\{\sigma_1, \sigma_3\}$, since $[K_{\{\sigma_1,\sigma_3\}} : \mathbb{Q}] = 2$. Clearly $\sigma_1(\alpha) + \sigma_3(\alpha)$ is left fixed by both σ_1 and σ_3, since $\{\sigma_1, \sigma_3\}$ is a group. We have

$$\sigma_1(\alpha) + \sigma_3(\alpha) = \alpha + \alpha^3 = i\sqrt{2}.$$

So $K_{\{\sigma_1,\sigma_3\}} = \mathbb{Q}[i\sqrt{2}]$. Similarly,

$$\sigma_1(\alpha) + \sigma_7(\alpha) = \alpha + \alpha^7 = \sqrt{2}$$

is left fixed by $\{\sigma_1, \sigma_7\}$. So $K_{\{\sigma_1,\sigma_7\}} = \mathbb{Q}[\sqrt{2}]$.

This technique is of no use in finding $K_{\{\sigma_1,\sigma_5\}}$, for

$$\sigma_1(\alpha) + \sigma_5(\alpha) = \alpha + \alpha^5 = 0.$$

But by a similar argument, $\sigma_1(\alpha)\sigma_5(\alpha)$ is left fixed by both σ_1 and σ_5, and

$$\sigma_1(\alpha)\sigma_5(\alpha) = \alpha\alpha^5 = -i.$$

Thus $K_{\{\sigma_1,\sigma_5\}} = \mathbb{Q}(-i) = \mathbb{Q}(i)$. $\square$

There are many proofs for the fundamental theorem of algebra. Next we will use the Galois theorem to give a very short algebraic proof for the fundamental theorem of algebra.

Theorem 6.2.1 (The Fundamental Theorem of Algebra). *The field of complex numbers $\mathbb{C}$ is algebraically closed and $\overline{\mathbb{R}} = \mathbb{C}$.*

Proof. We know that $\mathbb{C}$ is an algebraic extension field of $\mathbb{R}$ of degree 2. Let $p(x) \in \mathbb{C}[x]$ be irreducible. Any root u of $p(x)$ in the algebraic closure $\overline{\mathbb{C}}$ is algebraic over $\mathbb{R}$, so in $\mathbb{C}[x]$ we have $p(x) \mid \text{irr}(u, \mathbb{R})$.

The splitting field of $p(x)$ over $\mathbb{C}$ is contained in the splitting field E of $\mathrm{irr}(u, \mathbb{R}) \left(x^2 + 1 \right)$ over $\mathbb{R}$. The extension E over $\mathbb{R}$ is finite normal. Since $\mathbb{C} \leqslant E$, we have $2 \mid [E : \mathbb{R}]$ and so $2 \mid \mid \mathrm{Gal}(E/\mathbb{R}) \mid$.

Now consider a 2-Sylow subgroup $P \leqslant \mathrm{Gal}(E/\mathbb{R})$ ([ZTL, Theorem 3.4.1]). Then $\mid \mathrm{Gal}(E/\mathbb{R})\mid/\mid P\mid$ is odd. From Theorem 6.1.6, we have $P = \mathrm{Gal}(E/E_P)$ and

$$[E_P : \mathbb{R}] = \frac{\mid \mathrm{Gal}(E/\mathbb{R})\mid}{\mid \mathrm{Gal}(E/E_P)\mid} = \frac{\mid \mathrm{Gal}(E/\mathbb{R})\mid}{\mid P\mid}$$

which shows that $[E_P : \mathbb{R}]$ is odd. Theorem 5.4.14 allows us to write $E_P = \mathbb{R}(v)$ for some v whose minimal polynomial over $\mathbb{R}$ must also have odd degree. Since every real polynomial of odd degree has a real root, irreducibility implies that v has degree 1 over $\mathbb{R}$, and furthermore $E_P = \mathbb{R}$. Thus $\mathrm{Gal}(E/\mathbb{R}) = P$, i.e., $\mathrm{Gal}(E/\mathbb{R})$ is a 2-group.

Since $\mathrm{Gal}(E/\mathbb{C}) \leq \mathrm{Gal}(E/\mathbb{R})$, we know that $\mathrm{Gal}(E/\mathbb{C})$ is also a 2-group. There is a normal subgroup $N \lhd \mathrm{Gal}(E/\mathbb{C})$ of index 2 ([ZTL, Theorem 3.3.8]). We have the Galois extension E_N over $\mathbb{C}$ of degree 2. It is well-known that every quadratic $ax^2 + bx + c \in \mathbb{C}[x]$ has complex roots. So we cannot have an irreducible quadratic polynomial in $\mathbb{C}[x]$. We deduce that $\mid \mathrm{Gal}(E/\mathbb{C})\mid = 1$ and $E = \mathbb{C}$. Therefore any irreducible polynomial over $\mathbb{C}$ is of degree 1. Hence $\mathbb{C}$ is algebraically closed and $\overline{\mathbb{R}} = \mathbb{C}$. $\qquad\square$

6.3. Cyclotomic extensions.
This section deals with the splitting extension fields of $x^n - 1$ over a field F.

Definition 6.3.1. *The splitting field of $x^n - 1$ over F is called the nth* **cyclotomic extension** *of F.*

Lemma 6.3.2. *Let p be a prime and $n \in \mathbb{N}$ with $p \nmid n$, F a field of characteristic p. Then $x^n - 1$ has no multiple roots in $\overline{F}$.*

Proof. We see that $g(x) = (x^n - 1)' = nx^{n-1} \neq 0$. Then $\gcd(x^n - 1, g(x)) = 1$. By Theorem 4.2.10, we know that $x^n - 1$ has no multiple roots in $\overline{F}$. $\qquad\square$

Let us recall the Euler ϕ-function.

Definition 6.3.3. *The* **Euler ϕ-function** $\phi : \mathbb{N} \to \mathbb{N}$ *is defined by*

$$\phi(n) = \left| \left\{ k \in \{1, \cdots, n\} : \gcd(k, n) = 1 \right\} \right|.$$

Let $n = p_1^{r_1} p_2^{r_2} \cdots p_s^{r_s}$, where $p_1, \cdots, p_s$ are distinct primes, and $r_1, \cdots, r_s \in \mathbb{N}$. Then

$$\phi(n) = n \left(1 - \frac{1}{p_1}\right) \left(1 - \frac{1}{p_2}\right) \cdots \left(1 - \frac{1}{p_s}\right).$$

For example $\phi(20) = \phi(2^2 5) = 20(1 - 1/2)(1 - 1/5) = 8$.

Under the conditions in Lemma 6.3.2, let K be the splitting field of $x^n - 1$ over F. Then $x^n - 1$ has n distinct zeros in K, and these form a cyclic group of order n under the field multiplication. We saw that a cyclic group of order n has $\phi(n)$ generators. It is clear that these $\phi(n)$ generators are exactly the primitive nth roots of unity.

Definition 6.3.4. *Under the conditions in Lemma 6.3.2, the polynomial* $\Phi_n(x) = \prod_{i=1}^{\phi(n)} (x - \alpha_i)$, *where the* α_i *are all the primitive nth roots of unity in F, is called the nth* **cyclotomic polynomial** *over F.*

If $\text{char}(F) = p$ is a prime and $p|n$, then the nth cyclotomic polynomial over F is not well-defined at this moment since there is no primitive nth roots of unity. From Example 1.7.6 we know that $\Phi_n(x)$ can be reducible over a field of finite characteristic.

Since any automorphism of the Galois group $\text{Gal}(K/F)$ must permute the primitive nth roots of unity, we see that $\Phi_n(x)$ is left fixed under every element of $\text{Gal}(K/F)$ regarded as extended in the natural way to $K[x]$. Thus $\Phi_n(x) \in F[x]$. In particular, if $F = \mathbb{Q}$, then $\Phi_n(x) \in \mathbb{Q}[x]$, and $\Phi_n(x)|x^n - 1$. Thus over $\mathbb{Q}$, we must actually have $\Phi_n(x) \in \mathbb{Z}[x]$. Certainly we can consider this integer polynomial $\Phi_n(x)$ as a polynomial in $\mathbb{Z}_p[x]$ for any prime p. We have seen that, for any prime p, $\Phi_p(x)$ is irreducible over $\mathbb{Q}$ in Corollary 1.7.10. Actually all $\Phi_n(x)$ are irreducible over $\mathbb{Q}$.

Theorem 6.3.5. *For any $n \in \mathbb{N}$, $\Phi_n(x)$ is irreducible over $\mathbb{Q}$.*

Proof. Let ω be a primitive nth root of unity and let $f(x)$ be its irreducible polynomial over $\mathbb{Q}$. Since ω is also a zero of $x^n - 1$, it follows that $f(x)|x^n - 1$ and $f(x) \in \mathbb{Z}[x]$ by Gauss Lemma.

Claim 1: If $p \nmid n$ is a prime then ω^p is a zero of $f(x)$.

Suppose this claim is not true, i.e., ω^p is not a zero of $f(x)$. Let $g(x) = \text{irr}(\omega^p, \mathbb{Q})$. Then ω is a zero of $g(x^p)$. So $f(x)|g(x^p)$. Note that $f(x)g(x)|\Phi_n(x)$.

Now, considering $f(x)|g(x^p)$ in $\mathbb{Z}_p[x]$ we see that $f(x)|g(x)^p$ since $g(x^p) = g(x)^p$ in $\mathbb{Z}_p[x]$. Let $h(x)$ be an irreducible common factor of

$f(x)$ and $g(x)$ in $\mathbb{Z}_p[x]$. Then in $\mathbb{Z}_p[x]$, we have

$$h^2(x)|f(x)g(x), f(x)g(x)|x^n - 1,$$

i.e., $h^2(x)|x^n - 1$. This contradicts Lemma 6.3.2. The claim is now proved.

Using Claim 1 we see that $f(\omega^k) = 0$ for any integer k with $\gcd(k,n) = 1$. These ω^k are exactly all the primitive nth roots of unity. Thus $\Phi_n(x)|f(x)$. Hence $\Phi_n(x) = f(x)$. Thus $\Phi_n(x)$ is irreducible over $\mathbb{Q}$. □

Let us now limit our discussion to characteristic 0, in particular to subfields of the complex numbers. We know that $\cos(2\pi/n) + i\sin(2\pi/n)$ is a primitive nth root of unity, a zero of $\Phi_n(x)$.

Example 6.3.1. *Find the cyclotomic polynomial $\Phi_8(x)$ over $\mathbb{Q}$.*

Solution. A primitive 8th root of unity in $\mathbb{C}$ is

$$\omega = \cos(2\pi/8) + i\sin(2\pi/8) = 1/\sqrt{2} + i/\sqrt{2} = \frac{1+i}{\sqrt{2}}.$$

All the primitive 8th roots of unity in $\mathbb{Q}$ are $\omega, \omega^3, \omega^5$, and ω^7, so

$$\Phi_8(x) = (x - \omega)(x - \omega^3)(x - \omega^5)(x - \omega^7).$$

We can compute, directly from this expression, $\Phi_8(x) = x^4 + 1$.

We can also find $\Phi_8(x)$ from $\deg(\Phi_8(x)) = \phi(8) = 4$ and

$$x^8 - 1 = (x^4 + 1)(x^2 + 1)(x^2 - 1).$$

□

Theorem 6.3.6. *The Galois group $\mathrm{Gal}(K/\mathbb{Q})$ of the nth cyclotomic extension K of $\mathbb{Q}$ has $\phi(n)$ elements and is isomorphic to the abelian unit group $\mathcal{U}(\mathbb{Z}_n)$ of the commutative ring $(\mathbb{Z}_n, +, \cdot)$.*

Proof. Let

$$\omega = \cos(2\pi/n) + i\sin(2\pi/n).$$

So ω is a generator of the cyclic multiplicative group of order n consisting of all nth roots of unity. All the primitive nth roots of unity, that is, all the generators of this group, are of the form ω^m for $1 \le m < n$ with $\gcd(m,n) = 1$. The field $\mathbb{Q}(\omega)$ is the whole splitting field of $x^n - 1$ over $\mathbb{Q}$. Then $K = \mathbb{Q}(\omega)$. If ω^m is another primitive nth root of unity, then since ω and ω^m are conjugate over $\mathbb{Q}$, there is an automorphism τ_m in $\mathrm{Gal}(K/\mathbb{Q})$ mapping ω onto ω^m. Let τ_r be the similar automorphism in $\text{Gal}(K/\mathbb{Q})$ corresponding to a primitive nth root of unity ω^r. Then

$$(\tau_m \tau_r)(\omega) = \tau_m(\omega^r) = (\tau_m(\omega))^r = (\omega^m)^r = \omega^{rm}.$$

This shows that the Galois group $\text{Gal}(K/\mathbb{Q})$ is isomorphic to the group $\mathcal{U}(\mathbb{Z}_n)$ consisting of elements of $\mathbb{Z}_n$ relatively prime to n under multiplication modulo n. This group has $\phi(n)$ elements and is abelian. □

Using similar arguments as in the proof of Theorem 6.3.6 and the fact that $(\mathcal{R}_{x^n-1}, \cdot)$ is an abelian group (not necessarily of order $\phi(n)$) we can similarly deduce the following results.

Theorem 6.3.7. *Let E be the splitting field of $x^n - 1$ over F (not necessarily a cyclotomic extension). Then the Galois group $\text{Gal}(E/F)$ is abelian.*

Example 6.3.2. *For any prime p if $p|n$, prove that $\Phi_{np}(x) = \Phi_n(x^p)$ over $\mathbb{Q}$.*

Proof. We denote an nth primitive root of unity by ω_n. Since $p|n$ we see that $\phi(pn) = p\phi(n)$. Then $\Phi_{np}(x)$ and $\Phi_n(x^p)$ have the same degree $\phi(pn) = p\phi(n)$. Note that ω_{np}^p is an nth primitive root of unity. We see that ω_{np} is a root of $\Phi_n(x^p)$. Since $\Phi_{np}(x) = \text{irr}(\omega_{np}, \mathbb{Q})$, then $\Phi_{np}(x)|\Phi_n(x^p)$. Since both of them are monic and of the same degree therefore $\Phi_{np}(x) = \Phi_n(x^p)$. □

For some other properties of cyclotomic polynomials, see Sec. 6.7 (34–40).

Next we list the first few cyclotomic polynomials over $\mathbb{Q}$ of small composite degree.

$$\Phi_2(x) = x + 1$$
$$\Phi_4(x) = x^2 + 1$$
$$\Phi_6(x) = x^2 - x + 1$$
$$\Phi_8(x) = x^4 + 1$$
$$\Phi_9(x) = x^6 + x^3 + 1$$
$$\Phi_{10}(x) = x^4 - x^3 + x^2 - x + 1$$
$$\Phi_{12}(x) = x^4 - x^2 + 1$$
$$\Phi_{14}(x) = x^6 - x^5 + x^4 - x^3 + x^2 - x + 1$$

$$\Phi_{15}(x) = x^8 - x^7 + x^5 - x^4 + x^3 - x + 1$$
$$\Phi_{16}(x) = x^8 + 1$$
$$\Phi_{18}(x) = x^6 - x^3 + 1$$
$$\Phi_{20}(x) = x^8 - x^6 + x^4 - x^2 + 1$$
$$\Phi_{21}(x) = x^{12} - x^{11} + x^9 - x^8 + x^6 - x^4 + x^3 - x + 1$$
$$\Phi_{22}(x) = x^{10} - x^9 + x^8 - x^7 + x^6 - x^5 + x^4 - x^3 + x^2 - x + 1$$
$$\Phi_{24}(x) = x^8 - x^4 + 1$$
$$\Phi_{25}(x) = x^{20} + x^{15} + x^{10} + x^5 + 1$$
$$\Phi_{26}(x) = x^{12} - x^{11} + x^{10} - x^9 + x^8 - x^7$$
$$+ x^6 - x^5 + x^4 - x^3 + x^2 - x + 1$$
$$\Phi_{27}(x) = x^{18} + x^9 + 1$$
$$\Phi_{28}(x) = x^{12} - x^{10} + x^8 - x^6 + x^4 - x^2 + 1$$
$$\Phi_{30}(x) = x^8 + x^7 - x^5 - x^4 - x^3 + x + 1.$$

6.4. Solvability by radicals.

We knew that a quadratic polynomial $f(x) = ax^2 + bx + c, a \neq 0$, with real coefficients, has the following zeros in $\mathbb{C}$:

$$\frac{-b \pm \sqrt{b^2 - 4ac}}{2a}.$$

Actually, this formula holds for $f(x) \in F[x]$, where F is any field of characteristic $\neq 2$ and the zeros are in $\overline{F}$. For example, $x^2 + 2x + 3 \in \mathbb{Q}[x]$ has its zeros

$$\frac{-2 \pm \sqrt{-8}}{2} = -1 \pm \sqrt{-2} \in \mathbb{Q}(\sqrt{-2}).$$

You may wonder whether the zeros of a cubic polynomial over $\mathbb{Q}$ can also always be expressed in terms of radicals. The answer is yes indeed. Also the zeros of a polynomial of degree 4 (called quartic) over $\mathbb{Q}$ can be expressed in terms of radicals. But we will see that some 5th degree polynomial (called quintic) do not have the "radical formula" for their zeros. We will describe precisely what this means.

Definition 6.4.1. *An extension K of a field F is an* **extension of F by radicals** *(or* **a radical extension of F**) *if there are elements $\alpha_1, \cdots, \alpha_r \in K$ and positive integers $n_1, \cdots, n_r$ such that*

$K = F(\alpha_1, \cdots, \alpha_r), \alpha_1^{n_1} \in F$ and $\alpha_i^{n_i} \in F(\alpha_1, \cdots, \alpha_{i-1})$ *for* $1 < i \leq$ *r. We say an element* α *in an extension of* F *is* **expressible by radicals** *if* α *is contained in some radical extension of* F. *A polynomial* $f(x) \in F[x]$ *is* **solvable by radicals over** F *if the splitting field* $F_{f(x)}$ *of* $f(x)$ *over* F *is contained in a radical extension of* F.

Lemma 6.4.2. *Let* K *be a radical finite extension of a field* F. *Then there is a finite splitting radical extension* E *of* F *containing* K.

Proof. We need only to show the case where $K = F(\alpha)$ where $\alpha^n \in F$. Let $a = \alpha^n \in F$. Then $p(x) = \mathrm{irr}(\alpha, F) | x^n - a \in F[x]$. For any $\beta \in \mathcal{R}_{p(x)} = \{\alpha_1, \cdots, \alpha_r\}$ we also have $p(\beta) = 0 = \beta^n - a$ So $\beta^n = a \in F$. We see that $E = F(\alpha_1, \cdots, \alpha_r)$ is a finite splitting radical extension of F containing K since $\alpha_1^n, \cdots, \alpha_r^n \in F$. □

If further $\mathrm{char}(F) = 0$, in the above lemma E is a finite normal radical extension of F containing K.

Example 6.4.1. *Let* ω *be a primitive 5th root of unity. The splitting field of* $x^5 - 1$ *is* $K = \mathbb{Q}(\omega)$. *Thus the polynomial* $x^5 - 1$ *is solvable by radicals over* $\mathbb{Q}$.

Similarly, $x^5 - 2$ *is solvable by radicals over* K, *for its splitting field over* $\mathbb{Q}$ *is* $K(\sqrt[5]{2})$, *where* $\sqrt[5]{2}$ *is the real zero of* $x^5 - 2$. *Thus the polynomial* $x^5 - 2$ *is solvable by radicals over* $\mathbb{Q}$ *since* $\mathbb{Q} \leq \mathbb{Q}(\omega) \leq \mathbb{Q}(\omega, \sqrt[5]{2})$.

In this section we shall show that a polynomial $f(x) \in F[x]$ with $\mathrm{char}(F) = 0$ is solvable by radicals over a field F if and only if its splitting field E over F has a solvable Galois group $\mathrm{Gal}(E/F)$. Then we will find some quintic polynomials $f(x) \in \mathbb{Q}[x]$ with a splitting field E over $\mathbb{Q}$ such that $\mathrm{Gal}(E/\mathbb{Q}) \simeq S_5$, the symmetric group on 5 letters. Since S_5 is not solvable, the $f(x)$ is not solvable by radicals.

We shall start with an arbitrary field F in this section and later we shall restrict ourselves to fields of characteristic 0.

We first recall the definition for a group to be solvable and several related results from [ZTL].

Definition 6.4.3. *A group* G *is called* **solvable** *if it has a subnormal series whose factor groups (quotient groups) are all abelian, that is, if there are subgroups*

$$1 = G_0 < G_1 < \cdots < G_k = G$$

such that each G_{j-1} *is normal in* G_j, *and each* G_j/G_{j-1} *is an abelian group, for* $j = 1, 2, \cdots, k$.

Definition 6.4.4. *A subnormal series*

$$\{e\} = G_0 \lhd G_1 \lhd \cdots \lhd G_n = G$$

of a group G is a **composition series** *if all the factor groups G_{i+1}/G_i are simple.*

Theorem 6.4.5. *Let G be a group and N be a normal subgroup of G.*

(1). G is solvable iff both N and G/N are solvable.

(2). If G is solvable, and H is a subgroup of G, then H is solvable.

(3). If G and H are solvable, the direct product $G \times H$ is solvable.

(4). G is finite solvable if and only if G has a composition se-ries with all composition factors of prime order (which are certainly cyclic).

(5). The groups S_n and A_n are solvable if and only if $n \leq 4$.

Lemma 6.4.6. *Let $a \in F$. If K is the splitting field of $x^n - a$ over F, then $\mathrm{Gal}(K/F)$ is a solvable group.*

Proof. From Corollary 1.6.5 we know that the nth roots of unity form a cyclic subgroup $(U_n, \cdot)$ of $(\overline{F}^*, \cdot)$ with generator ω of order r. Then the nth roots of unity are

$$1, \omega, \omega^2, \cdots, \omega^{r-1}.$$

Let $\beta \in E$ be a zero of $x^n - a \in F[x]$. Then all zeros of $x^n - a$ are

$$\beta, \omega\beta, \omega^2\beta, \cdots, \omega^{r-1}\beta.$$

Case 1: $\omega \in F$.

In this case we have $K = F(\beta)$. Then an automorphism $\sigma \in \mathrm{Gal}(K/F)$ is determined by the value $\sigma(\beta)$. Now if $\sigma(\beta) = \omega^i \beta$ and $\tau(\beta) = \omega^j \beta$, where $\tau \in \mathrm{Gal}(K/F)$, then

$$(\tau\sigma)(\beta) = \tau(\sigma(\beta)) = \tau(\omega^i \beta) = \omega^i \tau(\beta) = \omega^i \omega^j \beta,$$

since $\omega^i \in F$. Similarly,

$$(\sigma\tau)(\beta) = \omega^j \omega^i \beta.$$

Thus $\sigma\tau = \tau\sigma$, and $\mathrm{Gal}(K/F)$ is abelian and therefore solvable.

Case 2: $\omega \notin F$, i.e., F does not contain any generator of $(U_n, \cdot)$.

Since $\beta, \omega\beta$ are both zeros of $x^n - a$, then $\omega = (\omega\beta)/\beta \in K$. Let $F' = F(\omega)$, so we have $F < F' \leq K$. Now F' is the splitting field of $x^n - 1$. Since $F' = F(\omega)$, an automorphism $\mu \in \mathrm{Gal}(F'/F)$ is determined by $\mu(\omega)$. We must have $\mu(\omega) = \omega^i$ for some i, since all zeros of $x^n - 1$ are powers of ω. If $\eta(\omega) = \omega^j$ for $\eta \in \mathrm{Gal}(F'/F)$, then

$$(\mu\eta)(\omega) = \omega^{ij}.$$

and, similarly,

$$(\eta\mu)(\omega) = \omega^{ij}.$$

Thus $\mathrm{Gal}(F'/F)$ is abelian. By Theorem 6.1.6,

$$\{\iota\} \leq \mathrm{Gal}(K/F') \leq \mathrm{Gal}(K/F)$$

is a normal series and hence a subnormal series of groups. The first part of the proof shows that $\mathrm{Gal}(K/F')$ is abelian, and Galois Theorem 6.1.6 tells us that

$$\mathrm{Gal}(K/F)/\mathrm{Gal}(K/F') \simeq \mathrm{Gal}(F'/F),$$

which is abelian. So $\mathrm{Gal}(K/F)$ is solvable by Theorem 6.4.5 (1). $\square$

Corollary 6.4.7. *Let F be a field with $\mathrm{char}(F) = 0$, E be a finite normal radical extension of F. Then $\mathrm{Gal}(E/F)$ is solvable.*

Proof. There exists a sequence of field extensions

$$F = F_0 \subset F_1 \subset \cdots \subset F_r = E$$

for which there exist $\beta_i \in F_i$ and positive integers n_i such that $F_i = F_{i-1}(\beta_i)$ and $\beta_i^{n_i} \in F_{i-1}$. Let $n = \mathrm{lcm}(n_1, \cdots, n_r)$ and let E' be the splitting field for $x^n - 1$ over E. Let ω be a generator of $(R_{x^n-1}, \cdot)$, the nth root of unity in E'. Then $E' = E(\omega)$ which is also a finite normal extension of F. Let $F' = F(\omega)$ and let $F_i' = F_i(\omega)$. Again we have a sequence of field extensions

$$F \subset F' = F_0' \subset F_1' \subset \cdots \subset F_r' = E'$$

such that $F_i' = F_{i-1}'(\beta_i)$ and $\beta_i^{n_i} \in F_{i-1}'$. Moreover F_i' is a normal extension of F_{i-1}' with $F_{-1}' = F$. Let $G_i = \mathrm{Gal}(E'/F_i')$. Then these form a chain of subgroups, all normal in $G_{-1} = \mathrm{Gal}(E'/F)$:

$$G_r = \{e\} \trianglelefteq G_{r-1} \trianglelefteq \cdots \trianglelefteq G_{-1}.$$

We see that

$$G_j/G_{j+1} = \mathrm{Gal}(E'/F_j')/\mathrm{Gal}(E'/F_{j+1}') \simeq \mathrm{Gal}(F_{j+1}'/F_j'),$$

$$G_{-1}/G_0 = \mathrm{Gal}(E'/F)/\mathrm{Gal}(E'/F_0') \simeq \mathrm{Gal}(F_0'/F).$$

Now $\mathrm{Gal}(F_{j+1}'/F_j')$ is solvable by Lemma 6.4.6. So each G_i is solvable for $i = r, r - 1, \cdots, -1$ by induction and using Theorem 6.4.5.

In particular, $G_{-1} = \mathrm{Gal}(E'/F)$ is solvable. Finally $\mathrm{Gal}(E/F) = \mathrm{Gal}(E'/F)/\mathrm{Gal}(E'/E)$ must also be solvable by Theorem 6.4.5 (3). $\qquad\square$

Let us recall the following definition from Sec. 6.1.

Definition 6.4.8. *Let $f(x) \in F[x]$. We define the **Galois group of** $f(x)$ to be the Galois group $\mathrm{Gal}(E/F)$ of a splitting field E of $f(x)$, denoted by $\mathrm{Gal}(f(x)/F)$.*

Let $\mathrm{Gal}(E/F)$ be the Galois group of a splitting field E of $f(x) \in F[x]$, which has been considered as a subgroup of the symmetric group on $R_{f(x)} = \{\alpha_1, \alpha_2, \cdots, \alpha_n\}$, all the distinct zeros of $f(x)$. See Lemma 6.1.7.

Theorem 6.4.9. *Let F be a field with $\mathrm{char}(F) = 0$ and $f(x) \in F[x]$. If $f(x)$ is solvable by radicals over F, then the Galois group $\mathrm{Gal}(f(x)/F)$ is solvable.*

Proof. Let E be the splitting field of $f(x)$ over F. Then E is normal over F since $\mathrm{char}(F) = 0$. Since $f(x)$ is solvable by radicals, by Lemma 6.4.2 we know that E is contained in a finite normal radical extension, say E'. Then $\mathrm{Gal}(E'/F)$ is solvable by Corollary 6.4.7. Therefore, $\mathrm{Gal}(E/F) = \mathrm{Gal}(E'/F)/\mathrm{Gal}(E'/E)$ is also solvable by Theorem 6.4.5 (3). $\qquad\square$

Next we will produce a concrete irreducible quintic polynomial whose Galois group is S_5.

Theorem 6.4.10. *Let p be a prime. Suppose that H is a subgroup of S_p that contains a transposition and a p-cycle. Then $H = S_p$.*

Proof. We may assume that $\sigma = (1,2), \tau_1 = (1, k_1, \cdots, k_{p-1}) \in H$. If $k_i = 2$ we see that $\tau_1^i = (1, 2, \cdots) \in H$. Without loss of generality we may assume that $\sigma = (1,2), \tau = (1, 2, \cdots, p) \in H$. Computing $\tau^i \sigma \tau^{-i}$ we see that

$$(1,2), \tau(1,2)\tau^{-1} = (2,3), \tau^2(1,2)\tau^{-2} = (3,4), \cdots, \tau^{p-1}(1,2)\tau^{1-p}$$
$$= (p-1, p) \in H.$$

Further

$$(1,2), (1,2)(2,3)(1,2) = (1,3), (1,3)(3,4)(1,3) = (1,4),$$
$$\cdots, (1, p-1)(p-1, p)(1, p-1) = (1, p) \in H.$$

Thus $H = S_p$. $\qquad\square$

Question: Can we change p to any $n \in \mathbb{N}$ in the previous theorem?

Theorem 6.4.11. *Let p be a prime, let $f(x) \in \mathbb{Q}[x]$ be an irreducible polynomial of degree p with exactly $p - 2$ real roots. Then the Galois group of $f(x)$ over $\mathbb{Q}$ is S_p. Consequently $f(x)$ is not solvable by radicals over $\mathbb{Q}$.*

Proof. Let α be a root of $f(x)$ and let E be its splitting field over $\mathbb{Q}$. Then E is finite normal over F, $[\mathbb{Q}(\alpha) : \mathbb{Q}] = p$ and $\mathbb{Q}(\alpha) \subset E$. So the order of $G = \mathrm{Gal}(E/\mathbb{Q})$ is divisible by p since

$$|G| = [E : \mathbb{Q}] = [E : \mathbb{Q}[\alpha]][\mathbb{Q}[\alpha] : \mathbb{Q}].$$

By the Sylow Theorem, we certainly know that G must contain an element τ of order p, which must be a p-cycle.

On the other hand the set of roots is invariant under complex conjugation, so complex conjugation permutes the roots of $f(x)$ and therefore restricts to an automorphism σ of E of order two. Note that we have identified G as a group of permutations of the roots. Then σ identifies with the transposition (α_1, α_2), where α_1, α_2 are the non-real roots. Hence G contains a transposition and p-cycle, so by the previous theorem $G = S_p$. $\qquad\square$

Question: Can we have the following result? "Let $f(x) \in \mathbb{Q}[x]$ be an irreducible polynomial of degree n with exactly $n - 2$ real roots. Then the Galois group of $f(x)$ over $\mathbb{Q}$ is S_n."

Example 6.4.2. *Show that $f(x) = 2x^5 - 10x + 5$ is irreducible over $\mathbb{Q}$, and find the number of real roots. Find the Galois group of $f(x)$ over $\mathbb{Q}$, and explain why $f(x)$ is not solvable by radicals.*

Solution. The polynomial $f(x)$ is irreducible over $\mathbb{Q}$ since it satisfies Schönemann-Eisenstein Criterion for $p = 5$. Consider $y = f(x)$ as a continuous real valued function. The derivative $f'(x) = 10x^4 - 10$ has two real roots ± 1. We see that

$$\begin{cases} f'(x) > 0 & \text{if } |x| > 1, \\ f'(x) < 0 & \text{if } |x| < 1. \end{cases}$$

That is, $f(x)$ is strictly increasing for $|x| > 1$, and $f(x)$ is strictly decreasing for $|x| < 1$. Since

$$f(-\infty) = -\infty, f(-1) = 13 > 0, f(1) = -3 < 0, f(+\infty) = +\infty,$$

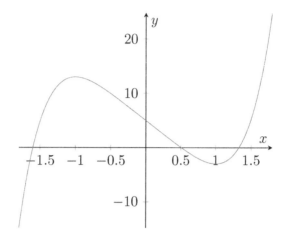

FIGURE 1. $y = 2x^5 - 10x + 5$

x	$-\infty$	$\nearrow$	-1	$\nearrow$	1	$\nearrow$	∞
$f'(x)$	∞	> 0	0	< 0	0	> 0	∞
$f(x)$	$-\infty$	$\nearrow$	13	$\searrow$	-3	$\nearrow$	∞

we see that $f(x)$ must have exactly three real roots. It follows from Theorem 6.4.11 that the Galois group of $f(x)$ over $\mathbb{Q}$ is S_5, and so it is not solvable. Thus $f(x)$ is not solvable by radicals. See Figure 1.
$\square$

We have finally proved the famous Abel-Ruffini Theorem which is named after Paolo Ruffini (1765–1822), who made an incomplete proof in 1799, and Niels Henrik Abel (1802–1829), who provided a proof in 1824. This is one of the greatest achievements of 19th century mathematics.

Theorem 6.4.12. *There exist quintic equations which are not solvable by radicals.*

As we mentioned earlier this immediately implies that no formula analogous to those for the cubic and quartic equations can exist for quintic equations.

To have perfect Galois theory let us prove the converse of Theorem 6.4.9 next.

Lemma 6.4.13. *Let E be a finite normal extension of F with cyclic $\mathrm{Gal}(E/F) = \langle \sigma \rangle$ of order $n > 1$, where F contains a primitive nth* *root of unity* ω. *Then* $\beta + \omega\sigma(\beta) + \cdots + \omega^{n-1}\sigma^{n-1}(\beta) \neq 0$ *for some* $\beta \in E$.

Proof. Since E be a finite normal extension of F then $E = F(\alpha)$ for some $\alpha \in E$. Then the numbers $\{\alpha, \sigma(\alpha), \ldots, \sigma^{n-1}(\alpha)\}$ are pairwise distinct. Let

$$g = 1 + \omega\sigma + \omega^2\sigma^2 + \cdots + \omega^{n-1}\sigma^{n-1}$$

which is an F-linear multiplicative function on E. To the contrary, we suppose $g(E) = 0$. Considering $g(1) = g(\alpha) = g(\alpha^2) = \cdots = g(\alpha^{n-1}) = 0$ we have

$$1 + \omega + \cdots + \omega^{n-1} = 0,$$

$$\alpha + \sigma(\alpha)\omega + \cdots + \sigma^{n-1}(\alpha)\omega^{n-1} = 0,$$

$$\alpha^2 + \sigma(\alpha)^2\omega + \cdots + \sigma^{n-1}(\alpha)^2\omega^{n-1} = 0,$$

$$\cdots\cdots\cdots$$

$$\alpha^{n-1} + \sigma(\alpha)^{n-1}\omega + \cdots + \sigma^{n-1}(\alpha)^{n-1}\omega^{n-1} = 0.$$

This is impossible since the coefficient matrix is the invertible Vandermonde matrix of $\alpha, \sigma(\alpha), \ldots, \sigma^{n-1}(\alpha)$. So $g(E) \neq 0$, the statement follows. $\qquad\square$

Theorem 6.4.14. *Let E be a finite normal extension of F with cyclic* $\mathrm{Gal}(E/F)$ *of order n, where F contains a primitive nth root of unity* ω. *Then there exists $a \in F$ such that $f(x) = x^n - a$ is irreducible over F and E is a splitting field for f over F. Consequently, E is a radical normal extension of F.*

Proof. Let $\mathrm{Gal}(E/F) = \langle\sigma\rangle$. From the previous lemma there is $\beta \in E$ such that $\theta = \beta + \omega\sigma(\beta) + \cdots + \omega^{n-1}\sigma^{n-1}(\beta) \neq 0$. Now

$$\sigma(\theta) = \sigma(\beta) + \omega\sigma^2(\beta) + \cdots + \omega^{n-2}\sigma^{n-1}(\beta) + \omega^{n-1}\sigma^n(\beta) = \omega^{-1}\theta$$

since $\sigma^n(\beta) = \beta$. We take $a = \theta^n$. Since $\mathrm{Gal}(E/F) = \langle\sigma\rangle$ and $\sigma(\theta^n) = (\sigma(\theta))^n = (\omega^{-1}\theta)^n = \theta^n$, then $\theta^n \in F$. Now by definition of a, θ is a root of $f(x) = x^n - a$, so the roots of $x^n - a$ are $\theta, \omega\theta, \cdots, \omega^{n-1}\theta$. Therefore $F(\theta)$ is a splitting field for $f(x)$ over F. Since $\sigma(\theta) = \omega^{-1}\theta$, the distinct automorphisms $1, \sigma, \cdots, \sigma^{n-1}$ can be restricted to distinct automorphisms of $F(\theta)$. Consequently, $n \leq |\mathrm{Gal}(F(\theta)/F)| = [F(\theta) : F] \leq \deg f = n$ so $[F(\theta) : F] = n$. It follows that $E = F(\theta)$ and (since f must be the irreducible polynomial of θ over F) f is irreducible over F. $\qquad\square$

In Theorem 6.4.14 we assumed that F contains a primitive nth root of unity ω. This cannot be satisfied if $\mathrm{char}(F) = p$ and $p|n$. In this case we have the following result. One can find a proof in [J].

Theorem 6.4.15. *Let* $\mathrm{char}(F) = p$, E *be a finite normal extension of* F *with* $\mathrm{Gal}(E/F)$ *of order* p. *Then there exists* $a \in E$ *such that* $E = F(a)$ *where* $a^p - a \in F$.

Theorem 6.4.16. *Let* F *be a field with* $\mathrm{char}(F) = 0$. *If* E *is a finite normal extension of* F *with solvable* $\mathrm{Gal}(E/F)$, *then* E *is contained in a finite radical normal extension of* F.

Proof. Since $\mathrm{Gal}(E/F)$ is solvable, by Theorem 6.4.5 there is a subnormal series

$$\{0\} = G_r \lhd \cdots \lhd G_1 \lhd G_0 = \mathrm{Gal}(E/F)$$

such that each G_k/G_{k+1} is cyclic. Let $E_k = E_{G_k}$. By Galois Theorem 6.1.6, we get a sequence of field extension

$$F = E_0 \subseteq \cdots \subseteq E_r = E.$$

We see that $G_k = \mathrm{Gal}(E/E_k)$. Moreover, we know that $E_k \subseteq E_{k+1}$ is a normal extension with Galois group $\mathrm{Gal}(E_{k+1}/E_k) \cong G_k/G_{k+1}$. So $\mathrm{Gal}(E_{k+1}/E_k)$ is cyclic. Let $|G_{k-1}/G_k| = n_k$, and $n = [E : F]$. Then $n = n_1 n_2 \cdots n_r$. Let ω be an nth primitive root of unity.

From Theorem 6.4.14 we know that if E_k contains a primitive n_kth root of unity, then E_{k+1} is a radical normal extension of E_k. However, we may not have this in our hands. Now we add ω to each field in the sequence:

$$
\begin{array}{ccccccc}
E_0(\omega) & \subseteq & E_1(\omega) & \subseteq \cdots \subseteq & E_r(\omega) & = E(\omega). \\
\cup| & & \cup| & & \cup| & \\
F = \quad E_0 & \subseteq & E_1 & \subseteq \cdots \subseteq & E_r & = E.
\end{array}
$$

From Theorem 6.4.14 we know that $E_0(\omega)$ is a normal radical extension of $E_0 = F$. We also know that $E_{k+1}(\omega)$ is a normal extension of $E_k(\omega)$ for all i. Next we show that $E_{k+1}(\omega)$ is a radical extension of $E_k(\omega)$ for all k.

We define a group homomorphism

$$\phi : \mathrm{Gal}(E_{k+1}(\omega)/E_k(\omega)) \to \mathrm{Gal}(E_{k+1}/E_k), \quad \sigma \mapsto \sigma|_{E_{k+1}}.$$

This is well-defined because E_{k+1} is normal extension of E_k, and hence $\sigma(E_{k+1}) = E_{k+1}$ for any $\sigma \in \mathrm{Gal}(E_{k+1}(\omega)/E_k(\omega))$.

Next we show that ϕ is injective. Let $\phi(\sigma) = \text{id}$. Since $\sigma \in$ $\text{Gal}(E_{k+1}(\omega)/E_k(\omega))$, it fixes $E_k(\omega)$. In particular, it fixes ω. So σ must fix the whole of $E_{k+1}(\omega)$. So $\sigma = \text{id}$.

By injectivity, we know that $\text{Gal}(E_{k+1}(\omega)/E_k(\omega))$ is isomorphic to a subgroup of $\text{Gal}(E_{k+1}/E_k)$. So $\text{Gal}(E_{k+1}(\omega)/E_k(\omega))$ is cyclic. By Theorem 6.4.14, we know that $E_{k+1}(\omega)$ is a radical normal extension of $E_k(\omega)$. Then $E_r(\omega) = E(\omega)$ is a finite radical normal extension of F containing E. $\qquad\square$

Corollary 6.4.17. *Let F be a field with $\text{char}(F) = 0$ and $f(x) \in$ $F[x]$. Then $f(x)$ can be solved by radicals if and only if the Galois group $\text{Gal}(f(x)/F)$ is solvable.*

Proof. ($\Rightarrow$) This is Theorem 6.4.9.

($\Leftarrow$) Let E be the splitting of $f(x)$ over F. Then E is a finite normal extension of F with solvable $\text{Gal}(E/F)$. From Theorem 6.4.16 we know that E is contained in a finite radical normal extension of F. So $f(x)$ is solvable by radicals. $\qquad\square$

Remark. When F has characteristic p, Corollary 6.4.17 fails. Actually Theorem 6.4.16 fails.

Now we have the explanations why a polynomial equation of degree ≤ 4 has radical solutions.

Corollary 6.4.18. *Let F be a field with $\text{char} F = 0$, and let $f(x) \in$ $F[x]$ with $\deg f(x) = n \leq 4$. Then $f(x)$ is solvable by radicals.*

Proof. From Lemma 6.1.7 we know that $\text{Gal}(f(x)/F)$ is isomorphic to a subgroup of S_n which is solvable when $n \leq 4$ by Theorem 6.4.5. Applying Corollary 6.4.17 we know that $f(x)$ is solvable by radicals. $\qquad\square$

6.5. Insolvability of equations of higher degree.

In this section we will construct some polynomials whose Galois group is S_n for any $n \in \mathbb{N}$.

Let F be a field, $E = F(x_1, \cdots, x_n)$, the field of rational functions over F in the variables $x_1, \cdots, x_n$. Then there is an injective homomorphism $S_n \to \text{Aut}_F(E)$ given by permutations of x_i.

We define the **field of symmetric rational functions** $K = E_{S_n}$ to be the fixed subfield of S_n in E. We first prove a few important results on symmetric rational functions.

Definition 6.5.1. *The* **elementary symmetric polynomials** *on n-variables* $x_1, \cdots, x_n$ *are* $s_1, s_2, \cdots, s_n$ *defined by*

$$s_i = \sum_{1 \leq k_1 < k_2 < \cdots < k_i \leq n} x_{k_1} \cdots x_{k_i}.$$

It is easy to see that

$$s_1 = x_1 + x_2 + \cdots + x_n,$$
$$s_2 = x_1 x_2 + x_1 x_3 + \cdots + x_{n-1} x_n,$$
$$\cdots\cdots\cdots$$
$$s_n = x_1 \cdots x_n.$$

(6.2)

Obviously, $s_1, \cdots, s_n \in K$.

Theorem 6.5.2. *Let* F *be a field,* $E = F(x_1, \cdots, x_n)$ *and* $K = E_{S_n}$.

(1) E *is the splitting field of* $f(x) = x^n - s_1 x^{n-1} + \cdots + (-1)^n s_n$ *over* K.

(2) $K = E_{S_n} \subseteq E$ *is a finite normal extension with* $\mathrm{Gal}(E/K) \simeq S_n$.

(3) $K = F(s_1, \cdots, s_n)$.

(4) $f(x)$ *is irreducible over* K.

Proof. (1) In $E[x]$, we have

$$f(x) = (x - x_1) \cdots (x - x_n).$$

So E is the splitting field of f over K.

(2) Since $f(x)$ is separable and E is the splitting field of $f(x)$, then $K = E_{S_n} \subseteq E$ is a finite normal extension. By Galois Theorem, we see that $\mathrm{Gal}(E/K) \simeq S_n$.

(3) Let $K_1 = F(s_1, \cdots, s_n)$. Clearly, $K_1 \subseteq K$. Now $K \subseteq E$ is a finite normal extension, since E is the splitting field of f over K and f has no repeated roots.

By Galois Theorem, since we have the finite normal extensions $K_1 \subseteq K \subseteq E$, we have $S_n = \mathrm{Gal}(E/K) \leq \mathrm{Gal}(E/K_1)$. We also know that $\mathrm{Gal}(E/K_1)$ is a subgroup of S_n, we must have $\mathrm{Gal}(E/K_1) = \mathrm{Gal}(E/K) = S_n$. So we must have $K = K_1$.

(4) If $f(x) = g(x)h(x)$ where $g(x), h(x) \in K[x]$ are of positive degree. For any $\sigma \in \mathrm{Gal}(E/K) = \mathrm{Gal}(f(x)/K)$ we would have

$$\sigma(\mathcal{R}_{g(x)}) = \mathcal{R}_{g(x)}, \quad \sigma(\mathcal{R}_{h(x)}) = \mathcal{R}_{h(x)},$$

which contradicts the fact that $\mathrm{Gal}(E/K) = S_n$. So $f(x)$ is irreducible over K. $\qquad\square$

Theorem 6.5.3. *Let F be a field with $\operatorname{char}F = 0$, $x_1, x_2, \cdots, x_n$ be variables over F, $K = F(s_1, s_2, \cdots, s_n) \leq F(x_1, \cdots, x_n)$. Then the polynomial*

$$f(x) = x^n - s_1 x^{n-1} + \cdots + (-1)^n s_n \in K[x]$$

is solvable by radicals over K if and only if $n < 5$.

Proof. We know that $K(x_1, x_2, \cdots, x_n)$ is the splitting field of the separable irreducible polynomial $f(x)$ over K. From Theorem 6.5.2 we know that $\operatorname{Gal}(E/K) = S_n$ which is not solvable if $n \geq 5$ by Theorem 6.4.5. From Theorem 6.4.9 we see that $f(x)$ is solvable by radicals over K if and only if $n < 5$. $\qquad\square$

6.6. Dedekind's Theorem and discriminants of polynomials.

In general it is difficult to compute the Galois groups of polynomials over a field, particularly for higher degree polynomials. We include the following theorem to help computing the Galois groups of polynomials over $\mathbb{Q}$. This theorem is named after Julius Wilhelm Richard Dedekind. We omit its proof here. For the detailed proof see Sec. 61 in [J].

Theorem 6.6.1 (Dedekind's Theorem). *Let $f(x) \in \mathbb{Z}[x]$ be monic with no repeated roots. Let $\overline{f}(x) \in \mathbb{Z}_p[x]$ be the obvious polynomial obtained by reducing the coefficients of $f \bmod p$ for a prime p. Assume that $\overline{f}(x)$ has no repeated roots. If $\overline{f}(x)$ factors as a product of irreducibles of degree $n_1, n_2, \cdots, n_r$, then $\operatorname{Gal}(f(x)/\mathbb{Q})$ contains an element of cycle type $[n_1, \cdots, n_r]$ (on $\mathcal{R}_{f(x)}$).*

Example 6.6.1. *Show that $f(x) = x^5 + 3x^2 + 2x + 3$ is not solvable by radicals over $\mathbb{Q}$.*

Solution. In $\mathbb{Z}_2[x]$, $f(x) = x^5 + x^2 + 1$ is irreducible.

In $\mathbb{Z}_3[x]$, $f(x) = x^5 - x = (x^2 + 1)x(x - 1)(x + 1)$ where $x^2 + 1$ is irreducible.

From Dedekind's Theorem we know that $\operatorname{Gal}(f(x)/\mathbb{Q}) \leq S_5$ contains a transposition and a 5-cycle. By Theorems 6.4.11 and 6.6.1 we see that $\operatorname{Gal}(f(x)/\mathbb{Q}) = S_5$ over $\mathbb{Q}$, and $f(x)$ is not solvable by radicals over $\mathbb{Q}$. $\qquad\square$

Discriminant of a polynomial is another tool to help computing the Galois group of a polynomial.

Definition 6.6.2 (Discriminant). *Let F be a field, $f(x) \in F[x]$, and E the splitting field of $f(x)$ over F with*

$$f(x) = a(x - \alpha_1) \cdots (x - \alpha_n), \quad a \in F, \alpha_1, \cdots, \alpha_n \in E.$$

We define

$$\Delta_{f(x)} = \prod_{i<j}(\alpha_i - \alpha_j), \quad D_{f(x)} = \Delta_{f(x)}^2 = (-1)^{n(n-1)/2} \prod_{i \neq j}(\alpha_i - \alpha_j),$$

*and call $D_{f(x)}$ the **discriminant** of f.*

Clearly, $D_{f(x)} \neq 0$ if and only if $f(x)$ has no repeated roots.

Theorem 6.6.3. *Let F be a field, $f(x) \in F[x]$, and E the splitting field of $f(x)$ over F. Suppose $D_{f(x)} \neq 0$ and $\mathrm{char}(F) \neq 2$. Then*

(1) $D_{f(x)} \in F$;

(2) $\mathrm{Gal}(E/F) \subseteq A_n$ if and only if $\Delta_{f(x)} \in F$ (if and only if $D_{f(x)}$ is a square in F).

Proof. (1). Note that E is a finite normal extension of F. It is clear that $D_{f(x)}$ is fixed by $\mathrm{Gal}(E/F)$ since it only permutes the roots.

(2). For any transposition $\sigma \in S_n$ switching α_i, α_j with $i \neq j$, we see that

$$\sigma(\Delta_{f(x)}) = -\Delta_{f(x)}.$$

So $\Delta_{f(x)} \in F$ if and only if $\Delta_{f(x)}$ is fixed by $\mathrm{Gal}(E/F)$ if and only if every element of $\mathrm{Gal}(E/F)$ is even, if and only if $\mathrm{Gal}(E/F) \leq A_n$. $\square$

Lemma 6.6.4. *Let F be a field with $\mathrm{char}(F) \neq 2, 3$. Let $f(x) = x^3 + bx + c \in F[x]$. Then $D_{f(x)} = -4b^3 - 27c^2$.*

Proof. Let $f(x) = (x - \alpha_1)(x - \alpha_2)(x - \alpha_2)$. Then

$$\alpha_1 + \alpha_2 + \alpha_3 = 0, \quad \alpha_1\alpha_2 + \alpha_1\alpha_3 + \alpha_2\alpha_3 = b, \quad \alpha_1\alpha_2\alpha_3 = -c,$$

i.e.,

$$b = -\alpha_1\alpha_2 - \alpha_1^2 - \alpha_2^2, \quad c = \alpha_1^2\alpha_2 + \alpha_2^2\alpha_1.$$

We compute

$$D_{f(x)} + 4b^3 + 27c^2 = (\alpha_1 - \alpha_2)^2(\alpha_1 - \alpha_3)^2(\alpha_3 - \alpha_2)^2 + 4b^3 + 27c^2$$

$$= (\alpha_1 - \alpha_2)^2(2\alpha_1 + \alpha_2)^2(2\alpha_2 + \alpha_1)^2 + 4(-\alpha_1\alpha_2 - \alpha_1^2 - \alpha_2^2)^3$$

$$+ 27(\alpha_1^2\alpha_2 + \alpha_2^2\alpha_1)^2 = 0$$

Thus $D_{f(x)} = -4b^3 - 27c^2$. $\square$

Corollary 6.6.5. *Let F be a field with* $\mathrm{char}(F) = 0$. *Let* $f(x) \in F[x]$ *be irreducible of degree 3. If $D_{f(x)}$ is a square of an element in F, then* $\mathrm{Gal}(E/F) = A_3$, *otherwise* $\mathrm{Gal}(E/F) = S_3$.

Proof. (1). If $D_{f(x)}$ is a square of an element in F, then $\Delta_{f(x)} \in F$, $\mathrm{Gal}(E/F) \subseteq A_3$ from Theorem 6.6.3. Note that $\mathrm{Gal}(E/F)$ has more that one element. So $\mathrm{Gal}(E/F) = A_3$.

(2). If $D_{f(x)}$ is not a square of an element in F, then $\Delta_{f(x)} \notin F$, $\mathrm{Gal}(E/F) \not\subseteq A_3$ from Theorem 6.6.3. The only nontrivial subgroups of S_3 are S_3, A_3 and $\{(1), (i,j)\}$ where $1 \leq i \leq j \leq 3$. Since $\mathrm{Gal}(E/F)$ acts on the three roots of $f(x)$ transitively, we must have $\mathrm{Gal}(E/F) = S_3$. □

By Theorem 6.1.8, the Galois group $\mathrm{Gal}(E/F)$ is always cyclic for any finite fields $F \leq E$.

Example 6.6.2. *Find* $\mathrm{Gal}(f(x)/\mathbb{Q})$ *for*
(a). $f(x) = x^3 - 2x + 2$,
(b). $f(x) = x^3 - 9x + 3$.

Solution.

(a). We know that $f(x)$ is irreducible over $\mathbb{Q}$ by Schönemann-Eisenstein criterion. Since $D_{f(x)} = -4(-2)^3 - 27(2)^2 < 0$, using Corollary 6.6.5 we know that $\mathrm{Gal}(f(x)/\mathbb{Q}) = S_3$.

(b). We know that $f(x)$ is irreducible over $\mathbb{Q}$ by Schönemann-Eisenstein Criterion. Since $D_{f(x)} = -4(-9)^3 - 27(3)^2 > 0$, using Corollary 6.6.5 we know that $\mathrm{Gal}(f(x)/\mathbb{Q}) = A_3$.

□

Similar to Lemma 6.6.4 we can obtain the following formula.

Lemma 6.6.6. *Let F be a field with* $\mathrm{char}(F) \neq 2, 3$. *Let* $f(x) = x^4 + cx^2 + dx + e \in F[x]$. *Then* $D_{f(x)} = 256e^3 - 128c^2e^2 + 144cd^2e - 27d^4 + 16c^4e - 4c^3d^2$. *In particular,* $D_{x^4+ax^2+b} = 16b(a^2 - 4b)^2$.

6.7. Exercises.

(1) Find the Galois groups of $x^4 - 2$ over the fields (a) $(\mathbb{Z}_3, +, \cdot)$, (b) $(\mathbb{Z}_7, +, \cdot)$.
(2) Find the Galois group of $x^4 + 2$ over the field (a) $(\mathbb{Z}_3, +, \cdot)$, (b) $(\mathbb{Z}_5, +, \cdot)$.
(3) Let $E \leq F$ be a finite normal extension. Let
$$G = \mathrm{Gal}(E/F), K = \{u \in E | \sigma\tau(u) = \tau\sigma(u) \forall \sigma, \tau \in G\}.$$

(a). Show that K is an intermediate subfield.

(b). Show that $F \leq K$ is a normal extension with abelian Galois group.

(4) Let ω be a primitive 20th root of unity in $\mathbb{C}$, and let $E = \mathbb{Q}(\omega)$.

    (a). Identify the Galois group $\mathrm{Gal}(E/\mathbb{Q})$, explaining how the individual automorphisms act on E.

    (b). How many subfields of E are there which are quadratic extensions of $\mathbb{Q}$?

    (c). Determine the irreducible polynomial of ω over $\mathbb{Q}$.

(5) Let K be the splitting field of $x^6 - 25$ over $\mathbb{Q}$. Determine $\mathrm{Gal}(K/\mathbb{Q})$. Explicitly determine all subfields of K, giving generators over $\mathbb{Q}$. Indicate which are normal over $\mathbb{Q}$.

(6) Let F be a field of characteristic 0, $F \subset E \subseteq F(x)$ be fields. Show that $\mathrm{Gal}(F(x)/E)$ is finite.

(7) Let F be a field of characteristic 0. Show that $F(x^2) \cap F(x^2 - x) = F$.

(8) Let F be a field of characteristic 0. Show that $F(x^2 - x) \cap F(x^{-1} - x) = F$.

(9) Let $F = \mathbb{Q}(\alpha)$, where $\alpha^3 = 5$. Determine the irreducible polynomial of $\alpha + \alpha^2$ over $\mathbb{Q}$.

(10) Consider $f(x) = x^4 + x^3 + x^2 + x + 1$ as a polynomial over $\mathbb{Q}$. How many subfields are there of the splitting field E of $f(x)$ over $\mathbb{Q}$? Justify your answer.

(11) Let E be a field containing exactly 64 elements. Find all the subfields of E. How many elements $\alpha \in E$ satisfy $E = \mathbb{Z}_2(\alpha)$? How many irreducible polynomials of degree 6 are there in $\mathbb{Z}_2[x]$?

(12) Let E be the splitting field of the function $f(x) = x^4 - 5$ over $\mathbb{Q}$. Find the Galois group G of $f(x)$ over $\mathbb{Q}$, and list all the subgroups of G and the corresponding fixed subfields.

(13) Let $E \leq L$ be a normal extension of fields, with Galois group $\mathrm{Gal}(L/E) = \{\sigma_1, \cdots, \sigma_n\}$, and let $\alpha \in L$. Show that $L = E(\alpha)$ if and only if $\sigma_1(\alpha), \cdots, \sigma_n(\alpha)$ are distinct.

(14) Let p be a prime. Demonstrate the existence of a normal extension of $\mathbb{Q}$ whose Galois group is cyclic with p elements.

(15) Let $f(x) \in F[x]$. Show that $f(x)$ is irreducible over F if and only if the action of $\mathrm{Gal}(f(x)/F)$ on $\mathcal{R}_{f(x)}$ is transitive.

(16) Let $f(x)$ be irreducible over $\mathbb{Q}$, and let F be its splitting field over $\mathbb{Q}$. Show that if $\mathrm{Gal}(F/\mathbb{Q})$ is abelian, then $F = \mathbb{Q}(u)$ for all roots u of $f(x)$.

(17) Let K be a field of characteristic 0 in which every cubic polynomial has a root. Let $f(x)$ be an irreducible quartic polynomial with coefficients in K whose discriminant is a square in K. What is the Galois group of $f(x)$ over K?

(18) Find the order of the Galois group of $x^5 - 2$ over $\mathbb{Q}$.

(19) Show that $f(x) = x^5 - 4x + 2$ is irreducible over $\mathbb{Q}$, and find the number of real roots. Find the Galois group of $f(x)$ over $\mathbb{Q}$, and explain why $f(x)$ is not solvable by radicals

(20) Calculate the Galois group of $x^3 - 3x + 1$ over $\mathbb{Q}$.

(21) Let $f(x) = x^3 - 3x - 1 \in \mathbb{Q}[x]$. Show that $\mathrm{Gal}(f(x)/\mathbb{Q}) = (\mathbb{Z}_3, +)$.

(22) Find infinitely many examples of polynomials of the form $x^3 + 2ax + a$ over $\mathbb{Q}$ with Galois group S_3.

(23) Calculate the Galois group of $x^4 + 5x + 5$ over $\mathbb{Q}$.

(24) Calculate the Galois group of $x^4 + px + p$ over $\mathbb{Q}$, where p is a prime greater than 5.

(25) Show that the Galois group of $f(x) = x^4 + 3x^2 + 2x + 1$ over $\mathbb{Q}$ is S_4. (Hint: Use Theorem 6.6.1.)

(26) Find infinitely many polynomials over $\mathbb{Q}$ with Galois group S_4. (Hint: Use Theorem 6.6.1 and the fact that S_4 is generated by a 4-cycle and a 3-cycle.)

(27) Show that $\mathbb{Q}(\sqrt{2 + \sqrt{2}})$ is normal over $\mathbb{Q}$ with Galois group isomorphic to the cyclic group $(\mathbb{Z}_4, +)$.

(28) Determine the Galois group of the polynomial $x^4 - 14x^2 + 9$ over $\mathbb{Q}$.

(29) Find the Galois groups of $x^3 - 2$ over the fields $\mathbb{Z}_5$, $\mathbb{Z}_7$ and $\mathbb{Z}_{11}$.

(30) Find the Galois group of $x^4 - 1$ over the field $\mathbb{Z}_7$.

(31) Let F be a finite, normal extension of $\mathbb{Q}$ for which $|\mathrm{Gal}(F/\mathbb{Q})| = 8$ and each element of $\mathrm{Gal}(F/\mathbb{Q})$ has order 2. Find the number of subfields of F that have degree 4 over $\mathbb{Q}$.

(32) Let $F = \mathbb{Q}(\sqrt{2}, \sqrt[3]{2})$. Find $[F : \mathbb{Q}]$ and prove that F is not normal over $\mathbb{Q}$.

(33) Find the Galois group of $x^9 - 1$ over $\mathbb{Q}$.

(34) For any prime p relatively prime to n, prove that $\Phi_{np}(x) = \frac{\Phi_n(x^p)}{\Phi_n(x)}$ over $\mathbb{Q}$.

(35) Prove that $\Phi_{12}(x^3) = \Phi_{18}(x^2)$ over $\mathbb{Q}$.

(36) For any prime p, show that $\Phi_{12}(x)$ is reducible in $\mathbb{Z}_p[x]$.

(37) Show that $\Phi_{10}(x)$ is irreducible in $\mathbb{Z}_2[x]$, but reducible in $\mathbb{Z}_5[x]$.

(38) Over $\mathbb{Q}$, show that

$$\prod_{d|n} \Phi_d(x) = x^n - 1.$$

(39) For any $n \in \mathbb{Z}^+$, show that $\Phi_{2^{n+1}}(x) = x^{2^n} + 1$ over $\mathbb{Q}$.

(40) For any $m, n \in \mathbb{N}$, show that $\Phi_n(x^m)$ over $\mathbb{Q}$ is irreducible over $\mathbb{Q}$ if and only if each prime factor of m is a factor of n.

(41) Let $n \in \mathbb{Z}^+$. Show that $f(x) = \left(x^2 + x\right)^{2^n} + 1$ is irreducible over $\mathbb{Q}$. (Hints: Show that $x^2 + x - \omega$ is irreducible over $\mathbb{Q}(\omega)$ where ω is a 2^{n+1}-th primitive root of unity.)

(42) Show that $x^4 - x^3 + x^2 - x + 1$ is irreducible over $\mathbb{Q}$, and use it to find the Galois group of $x^{10} - 1$ over $\mathbb{Q}$.

(43) Let $f(x) = x^5 - 5x^2 + 1$. Show $f(x)$ has precisely three real roots and is irreducible over $\mathbb{Q}$. Let $G = \text{Gal}(f(x)/\mathbb{Q})$, the Galois group of f over $\mathbb{Q}$. Show G contains a 5-cycle and a 2-cycle. What is the Galois group of $f(x)$? Is $f(x)$ solvable by radicals? explain.

(44) For any prime p, show that $f(x) = x^5 - p^2 x + p$ is irreducible over $\mathbb{Q}$, and find the number of real roots of $f(x)$. Find the Galois group of $f(x)$ over $\mathbb{Q}$, and explain why the group is not solvable. Consequently, $f(x)$ is not solvable by radicals.

(45) For any prime p, show that $f(x) = x^5 - 5p^4 x + p$ is irreducible over $\mathbb{Q}$, and find the number of real roots of $f(x)$. Find the Galois group of $f(x)$ over $\mathbb{Q}$, and explain why the group is not solvable. Consequently, $f(x)$ is not solvable by radicals.

(46) For any primes $q \geq p$ with $q \geq 5$, show that $f(x) = x^q - px + p$ is not solvable by radicals over $\mathbb{Q}$.

(47) Show that $f(x) = 15x^7 - 84x^5 - 35x^3 + 420x + 105$ is irreducible over $\mathbb{Q}$, and find the number of real roots of $f(x)$. Find the Galois group of $f(x)$ over $\mathbb{Q}$, and explain why the group is not solvable. Consequently, $f(x)$ is not solvable by radicals.

(48) (Kaplansky's Theorem). Let $f(x) = x^4 + ax^2 + b \in \mathbb{Q}[x]$ be irreducible.

(a). If b is a square in $\mathbb{Q}$ then $\text{Gal}(f(x)/\mathbb{Q}) \simeq \mathbb{Z}_2 \times \mathbb{Z}_2$.

(b). If $b(a^2 - 4b)$ is a square in $\mathbb{Q}$ then $\text{Gal}(f(x/\mathbb{Q}) \simeq \mathbb{Z}_4$.

(c). If neither b nor $b(a^2-4b)$ is a square in $\mathbb{Q}$ then $\text{Gal}(f(x/\mathbb{Q}) \simeq D_8$.

(Hints: You may assume that the roots of $f(x)$ are $\pm\alpha, \pm\beta$.)

(49) Show that $\mathbb{Z}[x_1, \cdots, x_n] \cap \mathbb{Q}(s_1, \cdots, s_n) = \mathbb{Z}[s_1, \cdots, s_n]$ where s_i are defined in (6.2).

(50) Let F be a field, $E = F(x_1, \cdots, x_n)$, $K = F(s_1, \cdots, s_n)$ where s_i are defined in (6.2).

(a). Find $[E : K]$, i.e., the dimension of the vector space E over the field K.

(b). Find a basis for the vector space E over the field K.

(51) Let G be an arbitrary finite group. Show that there is a field F and a polynomial $f(x) \in F[x]$ such that the Galois group of $f(x)$ is isomorphic to G. (Hint: Use Theorem 6.5.2.)

(52) Let $F \leq E$ be a finite normal extension. Let $\text{Gal}(E/F) = \{\varphi_1, \cdots, \varphi_n\}$. Define the **trace** and **norm** of $\alpha \in E$ as

$$\text{tr}_{E/F}(\alpha) = \sum_{i=1}^{n} \varphi_i(\alpha), \quad N_{E/F}(\alpha) = \prod_{i=1}^{n} \varphi_i(\alpha).$$

Show that $\text{tr}_{E/F}(\alpha), N_{E/F}(\alpha) \in F$.

(53) Let $F \leq E$ be a finite normal extension. Show that there is $\alpha \in E$ such that $\text{tr}_{E/F}(\alpha) \neq 0$.

(54) Let $F \leq E \leq K$ be finite normal extensions. Show that

$$\text{tr}_{K/F} = \text{tr}_{E/F} \circ \text{tr}_{K/E}, \quad N_{K/F} = N_{E/F} \circ N_{K/E}.$$

7. Sample Solutions

Chapter 1.

(1) Proof.

(a). Let $a \neq 0$. We want to show that a is not a 0-divisor. Note that $\varphi(a)$ is the unique element such that $a\varphi(a)a = a$. Suppose $ac = 0$ or $ca = 0$. Then $a(\varphi(a)+c)a = a\varphi(a)a+aca = a$. By uniqueness $\varphi(a) + c = \varphi(a)$ so $c = 0$.

(b). From $a\varphi(a)a = a$, we know that $\varphi(a) \neq 0$ also. Multiplying on the left by $\varphi(a)$ we obtain $\varphi(a)a\varphi(a)a = \varphi(a)a$. Because R has no divisors of zero by Part (a), multiplicative cancellation is valid and we see that $\varphi(a)a\varphi(a) = \varphi(a)$.

(c). We claim that $a\varphi(a)$ is unity for nonzero a and $\varphi(a)$ given in the statement of the exercise. Let $c \in R$. From $a\varphi(a)a = a$, we see that $ca = ca\varphi(a)a$. Canceling a, we obtain $c = c(a\varphi(a))$. From Part (b), we have $\varphi(a)c = \varphi(a)a\varphi(a)c$, and cancelling $\varphi(a)$ yields $c = (a\varphi(a))c$. Thus $a\varphi(a)$ satisfies $(a\varphi(a))c = c(a\varphi(a)) = c$ for all $c \in R$, so $a\varphi(a)$ is unity.

(d). Let a be a nonzero element of the ring. By Part (a), $a\varphi(a)a = a$. Using cancellation we obtain that $a\varphi(a) = 1$ and $\varphi(a)a = 1$. So $\varphi(a)$ is the inverse of a, and a is a unit. This shows that R is a division ring.

$\square$

(3) Solution.

(a). We have $y = y^6 = (-y)^6 = -y$, hence $2y = 0$ for any $y \in R$. Now let x be an arbitrary element in R. Using the binomial formula, we obtain

$$
\begin{aligned}
x + 1 = (x + 1)^6 \\
= x^6 + 6x^5 + 15x^4 + 20x^3 + 15x^2 + 6x + 1 \\
= x^4 + x^2 + x + 1,
\end{aligned}
$$

where we canceled the terms that had even coefficients. Hence $x^4 + x^2 = 0$, or $x^4 = -x^2 = x^2$. We then have

$$x = x^6 = x^2 x^4 = x^2 x^2 = x^4 = x^2,$$

and so $x^2 = x$, as desired.

(b). Expanding the equality $(x+y)^2 = x+y$ we deduce $xy+yx = 0$, so $xy = -yx = yx$ for any $x, y \in R$. This shows that the ring is commutative, as desired.

$\square$

(5) **Proof.** Let $G = R \setminus \{0\}$. We need to prove that $(G, \cdot)$ is a group, i.e., the identity axiom and the inverse element axiom hold for $(G, \cdot)$.

Let $a, b, c \in G$. Because of the cancellation rules we see that, $ab = ac$ iff $b = c$. Thus we see that $aG = G$. Since $a \in G = aG$, there exists $e_a \in G$ so that $ae_a = a$.

From this we also have $ae_a a = aa$. By the cancellation rules we have $e_a a = a$.

From $ab = ae_a b$ and the cancellation rules we have $b = e_a b$. Thus $e = e_a$ is the identity element in G.

Since $e \in G = aG = Ga$, there exists $x, y \in G$ so that $ax = ya = e$. Then $x = ex = yax = ye = y$ is the inverse of a. Therefore G is a group.

$\square$

(6) **Proof.** We need only to show that "+" is commutative.

Denote by 1 the identity element of $\langle S^*, \cdot \rangle$. For any $x \in S$, using (iii) we have $0x = x0 = 0$ and $x1 = 1x = x$.

For any $x, y \in S$, using (iii) we have

$$(1+x)(1+y) = 1+y+x+xy, \text{ and } (1+x)(1+y) = 1+y+x+xy.$$

Thus $x + y = y + x$, i.e., "+" is commutative.

Therefore, $\langle S, +, \cdot \rangle$ is a division ring.

$\square$

(8) **Solution.** The Polynomials of degree 3 in $\mathbb{Z}_2[x]$ are

x^3: not irreducible because 0 is a zero,

$x^3 + 1$: not irreducible because 1 is a zero,

$x^3 + x$: not irreducible because 0 is a zero,

$x^3 + x^2$: not irreducible because 0 is a zero,

$x^3 + x + 1$: irreducible, neither 0 nor 1 is a zero,

$x^3 + x^2 + 1$: irreducible, neither 0 nor 1 is a zero,

$x^3 + x^2 + x$: not irreducible because 0 is a zero,

$x^3 + x^2 + x + 1$: not irreducible, 1 is a zero.

Thus the irreducible cubics are $x^3 + x + 1$ and $x^3 + x^2 + 1$. $\square$

Solution 2. An irreducible polynomials of degree 3 in $\mathbb{Z}_2[x]$ has to be of the form $f(x) = x^3 + ax^2 + bx + 1$ for some $a, b \in \mathbb{Z}_2$. We know that f is irreducible iff $f(0)f(1) \neq 0$, iff $a + b \neq 0$, iff $a = 1$ and $b = 0$, or $a = 0$ and $b = 1$. Thus the irreducible cubics are $x^3 + x + 1$ and $x^3 + x^2 + 1$. □

(10) **Solution.** First assume that $1 - ab$ is invertible, and let $x = (1 - ab)^{-1}$. Then $x(1 - ab) = 1$, so $bx(1 - ab)a = ba$, and therefore $(1 - ba) + bx(1 - ab)a = 1$. Now

$$1 = (1 - ba) + bx(a - aba) = (1 - ba) + bxa(1 - ba) = (1 + bxa)(1 - ba).$$

It can be checked easily that $(1 - ba)(1 + bxa) = 1$, so $(1 - ba)^{-1} = (1 + bxa)$. A similar argument shows that if $1 - ba$ is invertible, then so is $1 - ab$. □

(11) **Solution.** We know that all possible rational zeros of the polynomial are $\pm 1, \pm 2, \pm 1/2$. By direct verification we see that the only rational zeros of the polynomial are $-2, 1/2$. By simple computations we see that $2x^4 + 3x^3 + 3x - 2 = (x + 2)(2x - 1)(x^2 + 1)$. Thus all real roots of the polynomial are $-2, 1/2$. □

(12) **Solution.** $f(x) = x(x^4 - 2x^2 + x + 2) = x(x + 1)(x^3 - x^2 - x + 2)$. It is easy to check that $x^3 - x^2 - x + 2$ does not have any solutions in $\mathbb{Z}_5$. Thus it is irreducible. So the irreducible factorization is $f(x) = x(x + 1)(x^3 - x^2 - x + 2)$. □

(13) **Solution.** It is irreducible. If $x^3 + 3x^2 - 8$ is reducible over $\mathbb{Q}$, then it factors in $\mathbb{Z}[x]$, and must therefore have a linear factor of the form $x - a$ in $\mathbb{Z}[x]$. Then a must be a zero of the polynomial and must divide -8, so the possibilities are $a = \pm 1, \pm 2, \pm 4, \pm 8$. Computing the polynomial at these eight values, we find none of them is a zero of the polynomial, which is therefore irreducible over $\mathbb{Q}$. □

(14) **Proof.** It is clear that ϕ is a ring homomorphism. Similarly $\psi : R \to R$ defined by $\psi(f(x)) = f(a^{-1}x - a^{-1}b)$ is also a ring homomorphism. It is easy to verify that $\psi = \phi^{-1}$. Thus ϕ is an automorphism of $R[x]$. □

(15) **Proof.** Since $\langle n, x \rangle = \mathbb{Z}[x]x + n\mathbb{Z}$ and $\langle x \rangle = \mathbb{Z}[x]x$, we see that

$$\frac{\mathbb{Z}[x]}{\mathbb{Z}[x]x} \simeq \mathbb{Z}, \quad \frac{\mathbb{Z}[x]x + n\mathbb{Z}}{\mathbb{Z}[x]x} \simeq n\mathbb{Z}.$$

Then

$$\mathbb{Z}[x]/\langle n, x\rangle \simeq \frac{\frac{\mathbb{Z}[x]}{\mathbb{Z}[x]x}}{\frac{\mathbb{Z}[x]x+n\mathbb{Z}}{\mathbb{Z}[x]x}} \simeq \frac{\mathbb{Z}}{n\mathbb{Z}} \simeq \mathbb{Z}_n.$$

So $\langle n, x\rangle$ is a prime ideal iff $\mathbb{Z}_n$ is an integral domain, which occurs iff n is a prime number.   □

(18) **Proof.** We first note that R is an integral domain since the zero ideal is prime. Let a be a nonzero element of R. If $a^2R = R$, we see that $a \in \mathcal{U}(R)$. Otherwise, by assumption, the ideal a^2R is prime, and so $a^2 \in a^2R$ implies $a \in a^2R$. Thus $a = a^2r$ for some $r \in R$, and since R is an integral domain, we can cancel a to obtain $1 = ar$, showing that a is invertible, which is impossible. So R is a field.   □

(26) **Solution.** In $\mathbb{Z}_5[x]$ we have $x^2 - 3 = x^2 + 2$. Then $R_1 = R_2$. In $\mathbb{Z}_2[x]$ we have $R_1 = \mathbb{Z}_2[x]/((x+1)^2)$, $R_2 = \mathbb{Z}_2[x]/(x^2)$. Then $R_1 \simeq R_2$ by the map $f : R_1 \to R_2, f(ax + b) = a(x - 1) + b$.

In $\mathbb{Z}_{11}[x]$ since $(2x)^2 - 3 = 4(x^2 + 2)$ then $R_1 \simeq R_2$ by the map $f : R_1 \to R_2, f(ax + b) = 2ax + b$.   □

(33) **Solution.**

(a). Let $a, b \in \sqrt{A}$. Then $a^m \in A$ and $b^n \in N$ for some positive integers m and n. In a commutative ring, the binomial expansion is valid. Consider $(a+b)^{m+n}$. In the binomial expansion, each summand contains a term $a^i b^{m+n-i}$. Now either $i > m$ so that $a^i \in A$ or $m + n - i > n$ so that $b^{m+n-i} \in A$. Thus each summand of $(a + b)^{m+n}$ is in A, so $(a + b)^{m+n} \in A$ and $a + b \in \sqrt{A}$. Also $sa \in \sqrt{A}$ since $(as)^m, (sa)^m \in A$ for any $s \in R$. Because $0^1 \in A$, we see that $0 \in \sqrt{A}$. Also $(-a)^m$ is either a^m or $-a^m$, and both a^m and $-(a^m)$ are in A. Thus $-a \in \sqrt{A}$. This shows that $\sqrt{A}$ is an ideal of R.

(b). Since $\sqrt{A} + \sqrt{B} = R$, there exist $a \in \sqrt{A}$ and $b \in \sqrt{B}$ such that $a + b = 1$. There exists an integer n such that $a^n \in A$ and $b^n \in B$. Then

$$1 = (a + b)^{2n} = \sum_{i=0}^{n} \binom{2n}{i} a^{2n-i} b^i + \sum_{i=1}^{n} \binom{2n}{n+i} a^{n-i} b^{n+i}$$

$$= a^n \sum_{i=0}^{n} \binom{2n}{i} a^{n-i} b^i + b^n \sum_{i=1}^{n} \binom{2n}{n+i} a^{n-i} b^i.$$

Clearly $x = a^n \sum_{i=0}^{n} \binom{2n}{i} a^{n-i} b^i \in A$ and $y = b^n \sum_{i=1}^{n} \binom{2n}{n+i} a^{n-i} b^i \in B$. From $x + y = 1$ we see that $A + B = R$.

□

(35) **Solution.** We know that $E = 2\mathbb{Z}$. Take $M = 4\mathbb{Z}$. Then $M \trianglelefteq E$, and $E/M = \{0 + 4\mathbb{Z}, 2 + 4\mathbb{Z}\}$. It is easy to see that M is a maximal ideal. Since $(2 + 4\mathbb{Z})^2 = 0$ therefore E/M is not a field. □

(36) **Solution.** Let $\mathcal{I}$ be the set of all prime ideals of R, $P = \bigcap_{I \in \mathcal{I}} I$, N is the set of all nilpotent elements of R. We will show that $N = P$.

In an example in class we know that $N \subset I$ for any $I \in \mathcal{I}$, yielding that $N \subset P$.

It is enough to show that for any $a \in R \setminus N$ there is $I \in \mathcal{I}$ with $a \notin I$. Let $A = \{a^k : k \in \mathbb{Z}_+\}$. Using Kuratowski-Zorn Lemma there is an ideal $M \trianglelefteq R$ such that M is maximal $A \cap M = \{0\}$. We shall show that M is prime.

Let $x, y \in R \setminus M$ such that $xy \in M$. We will try to deduce some contradictions. Consider the ideals $M + Rx$ and $M + Ry$ which properly contain M. Then $(M + Rx) \cap A \neq \{0\}$ and $(M + Ry) \cap A \neq \{0\}$. Thus there are $k_1, k_2 \in \mathbb{Z}_+, m_1, m_2 \in M, r_1, r_2 \in R$ such that

$$a^{k_1} = m_1 + r_1 x, a^{k_2} = m_2 + r_2 y.$$

We deduce that

$$a^{k_1+k_2} = (m_1 + r_1 x)(m_2 + r_2 y) \in M$$

a contradiction. Thus M is prime. □

(37) **Solution.** Let I be a maximal ideal of $\mathbb{C}[x]$. Then we know that $I = \langle f(x) \rangle$ for some irreducible $f(x) \in \mathbb{C}[x]$. We know that all irreducible polynomials are of degree 1. Then $f(x) = ax - b$ for some $a, b \in \mathbb{C}$ with $a \neq 0$. So $I = \langle ax + b \rangle = \langle x - c \rangle$ for some $c \in \mathbb{C}$. □

(55) **Answer.** $(x^4 + x^3 + x^2 + x + 1)(x^8 - x^7 + x^5 - x^4 + x^3 - x + 1)$.

Chapter 2.

(2) **Proof.** Let $a, b \in D^*$ that $p|ab$. Then there is $c \in D$ such that $ab = pc$. We may assume that

$$a = p_1 p_2 \cdots p_{s_1}, \quad b = q_1 q_2 \cdots q_{s_2}, \quad c = r_1 r_2 \cdots r_{s_3},$$

where p_i, q_j, r_k are irreducibles in D. Then

$$p_1 p_2 \cdots p_{s_1} q_1 q_2 \cdots q_{s_2} = p r_1 r_2 \cdots r_{s_3}.$$

Since D is an UFD, we see that $p \sim p_i$ or $p \sim q_j$. Thus $p|a$ or $p|b$, i.e., p is prime. $\qquad \square$

(4) **Answer.** $x^3 + 2x - 1$.

(8) **Solution.** Since $\frac{15-12i}{6-5i} = \frac{150}{61} + \frac{3i}{61}$, we have $15 - 12i = 2(6 - 5i) - (3 - 2i)$. Since $\frac{6-5i}{3-2i} = \frac{28}{13} - \frac{3i}{13}$, we have $6 - 5i = 2(3 - 2i) - i$. We put them together

$$15 - 12i = 2(6 - 5i) - (3 - 2i)$$
$$6 - 5i = 2(3 - 2i) - i$$
$$3 - 2i = i(-2 - 3i) + 0.$$

Thus $\gcd(15 - 12i, 6 - 5i) \sim 1$.

Since $\frac{16+7i}{10-5i} = 1 + \frac{6}{5}i$, we have $16 + 7i = (10 - 5i)(1 + i) + (1 + 2i)$. Since $\frac{10-5i}{1+2i} = -5i$, we have $10 - 5i = (1 + 2i)(-5i)$. We put them together

$$16 + 7i = (10 - 5i)(1 + i) + (1 + 2i)$$
$$10 - 5i = (1 + 2i)(-5i) + 0.$$

Thus we have $\gcd(16 + 7i, 10 - 5i) \sim 1 + 2i$. $\qquad \square$

(13) **Solution.** We work on the UFD $\mathbb{Z}[\sqrt{-2}][x, y]$. Let $v(a + b\sqrt{-2}) = a^2 + 2b^2$ be the standard multiplicative norm on $\mathbb{Z}[\sqrt{-2}]$. Write the equation as $(x + \sqrt{-2})(x - \sqrt{-2}) = y^3$.

(a). Since $v(\sqrt{-2}) = 2$, by Theorem 2.5.6 we know that $\sqrt{-2}$ is prime in $\mathbb{Z}[\sqrt{-2}]$.

(b). Let p be a prime in $\mathbb{Z}[\sqrt{-2}]$, $p|x + \sqrt{-2}$ and $p|x - \sqrt{-2}$. Then $p|2\sqrt{-2} = (\sqrt{-2})^3$. We know that $p = \pm\sqrt{-2}$. Then $\sqrt{-2}|x$, and $2|x^2$, and furthermore $4|x^2$, which is impossible from $x^2 + 2 = y^3$ (we would have $2^3|y^3 = x^2 + 2$). So $(x + \sqrt{-2})$, $(x - \sqrt{-2})$ are relatively prime.

(c). Since $(x + \sqrt{-2})$, $(x - \sqrt{-2})$ are relatively prime, from $(x + \sqrt{-2})(x - \sqrt{-2}) = y^3$ we know that $x + \sqrt{-2} = y_1^3$ for some $y_1 \in \mathbb{Z}[\sqrt{-2}]$. Let $y_1 = a + b\sqrt{-2}$. We have

$$x + \sqrt{-2} = (a + b\sqrt{-2})^3 = a^3 - 6ab^2 + (3a^2b - 2b^3)\sqrt{-2},$$

yielding that

$$1 = 3a^2b - 2b^3 = b(3a^2 - 2), \quad x = a^3 - 6ab^2.$$

We obtain that $b = 1$ and $a = \pm 1$, hence $x = \pm 5$ and $y = 3$.

$\qquad \square$

(16) **Solution.** Since R is a PID, we may suppose that $I = \langle a \rangle$, $J = \langle b \rangle$ for some nonzero $a, b \in D$. Let $d = \gcd(a, b)$. Then $IJ = \langle ab \rangle$, $I + J = \{ax + by | a, b \in R\} = \langle d \rangle$.

It is clear that $\langle ab/d \rangle \subset I \cap J$.

Let $z \in D$. Then $z \in I \cap J$ iff $z = ax = by$ for some $x, y \in R$, iff $a | by$, $\frac{a}{d} | \frac{b}{d} y$, iff $\frac{a}{d} | y$, iff $b \frac{a}{d} | by$ iff $b \frac{a}{d} | z$. Thus $z \in \langle ab/d \rangle$, i.e., $I \cap J \subseteq \langle ab/d \rangle$, and further $I \cap J = \langle ab/d \rangle$.

We see that $IJ = I \cap J$ iff $\langle ab/d \rangle = \langle ab \rangle$ iff d is invertile, iff $I + J = R$. □

(17) **Solution.**

(a). We see that $R = \{a_0 + a_2 x^2 + a_3 x^3 + \cdots a_n x^n | n \in \mathbb{Z}_+, a_i \in F\}$. Clearly R is a subring of $F[x]$.

(b). Suppose that $x^2 = fg$ for some $f, g \in R$. By comparing the degree of f and g we deduce that $\deg(f) = 0$ or $\deg(g) = 0$, that is, f or g is in $\mathcal{U}(R) = F^*$. So x^2 is irreducible in R. Similarly, x^3 is irreducible in R as well.

(c). From $(x^2)^3 = (x^3)^2$, we deduce that $x^2 | (x^3)^2$, but $x^2 \nmid x^3$, and that $x^2 | (x^3)^2$, but $x^2 \nmid x^3$ in R. We know that both x^2 and x^3 are irreducibles but not primes. So R is not a UFD.

(d). We claim that the ideal $\langle x^2, x^3 \rangle$ of R is not a principal ideal. Otherwise suppose that $\langle x^2, x^3 \rangle = \langle f(x) \rangle$ for some $f \in R$. It is clear that $\langle x^2, x^3 \rangle \neq R$. So f is not a unit. There are $u, v \in R$ such that $x^2 = uf, x^3 = vf$. Since x^2, x^3 are irreducible, we deduce that $u, v \in \mathcal{U}(R)$. Thus $x^2 \sim x^3$ which is impossible. Our claim follows.

□

(18) **Solution.** Let $D = \mathbb{C}[x]$ which is clearly a UFD. We see that $\mathbb{C}[x, y] = D[y]$.

Note that $y^5 + xy^4 - y^4 + x^2 y^2 - 2xy^2 + y^2 + x^3 - 1 = y^5 + (x - 1)y^4 + (x - 1)^2 y^2 + x^3 - 1$. Clearly $x - 1$ is irreducible in D. Taking $p = x - 1$ in Schönemann-Eisenstein Criterion we know that $y^5 + xy^4 - y^4 + x^2 y^2 - 2xy^2 + y^2 + x^3 - 1$ is irreducible in $\mathbb{C}[x, y]$.

Let $f(x, y) = xy^3 + x^2 y^2 - x^5 y + x^2 + 1$ which has degree 3 in y. Consider $y^3 f(x, 1/y) = x + x^2 y - x^5 y^2 + (x^2 + 1)y^3$. Taking $p = x$ in Schönemann-Eisenstein Criterion we know that $y^3 f(x, 1/y)$ is irreducible in the integral domain $\mathbb{C}[x, y]$. Thus $xy^3 + x^2 y^2 - x^5 y + x^2 + 1$ is irreducible in the integral domain $\mathbb{C}[x, y]$. □

(21) **Solution.** Let $d = 2, 3$. Note that if we define $v(a + b\sqrt{d}) = a^2 - db^2$, v is not a norm (since it can be negative). So we define

$v(a + b\sqrt{d}) = |a^2 - db^2|$. This is clearly a non-negative integer. Moreover, since d is not a square of an integer, $a^2 - db^2 = 0$ if $a = b = 0$. So $v(\alpha) > 0$ if $\alpha \neq 0$.

It is easy to see that $v(\alpha\beta) = v(\alpha)v(\beta) \leq v(\alpha)v(\beta)$ for nonzero $\alpha, \beta \in \mathbb{Z}[\sqrt{d}]$. We can obviously extend v to $\mathbb{Q}[\sqrt{d}]$ such that $v(\alpha\beta) = v(\alpha)v(\beta)$ for $\alpha, \beta \in \mathbb{Q}[\sqrt{d}]$.

For any $\alpha, \beta \in \mathbb{Z}[\sqrt{d}]$ with $\beta \neq 0$, let $\alpha/\beta = a + b\sqrt{d}$ where $a, b \in \mathbb{Q}$. Take $q = x + y\sqrt{d}$ where $x, y \in \mathbb{Z}$ with $|a - x| \leq 1/2$, $|b - y| \leq 1/2$. Let $r = \alpha - q\beta \in \mathbb{Z}[\sqrt{d}]$. We know that $\alpha = q\beta + r$, and

$$v(r) = v(\alpha - q\beta) = v(\beta)v(\alpha/\beta - q)$$
$$= v(\beta)|(a - x)^2 - (b - y)^2 d| \leq 3v(\beta)/4 < v(\beta).$$

$\square$

(24) **Solution.** Suppose that $f(x) = g(x)h(x)$ for some $g(x)$, $h(x) \in \mathbb{Z}[x]$ with $\deg(f(x)) > 0$ and $\deg(g(x)) > 0$. We may further assume that $g(x), h(x)$ are monic. We see that $g(a_i)h(a_i) = -1$, and furthermore $g(a_i) + h(a_i) = 0$. Since $\deg(g(x) + h(x)) \leq n - 1$, we deduce that $g(x) + h(x) = 0$, i.e., $h(x) = -g(x)$ which is impossible.

$\square$

(25) **Solution.** Suppose that $f(x) = g(x)h(x)$ for some $g(x)$, $h(x) \in \mathbb{Z}[x]$ with $\deg(f(x)) > 0$ and $\deg(g(x)) > 0$. We may further assume that $g(x), h(x)$ are monic. Both $g(x)$ and $h(x)$ do not have any real zeros. So $g(a_i) > 0$ and $h(a_i) > 0$ for all i. We see that $g(a_i)h(a_i) = 1$, and furthermore $g(a_i) = h(a_i) = 1$ for all i. If $\deg(g(x)) < n$, we deduce that $g(x) = 1$ which is impossible. If $\deg(h(x)) < n$, we deduce that $h(x) = 1$ which is impossible. Now we come to the case that $\deg(g(x)) = \deg(h(x)) = n$. Then $\deg(g(x) - h(x)) < n$, yielding that $g(x) - h(x) = 0$ since $g(a_i) - h(a_i) = 0$ for all i. You see the contradictions.

$\square$

(27) **Solution.** Suppose that $g(x) = g_1(x)g_2(x)$ for some $g_1(x)$, $g_2(x) \in \mathbb{Z}[x]$ with $\deg(g_1(x)) > 0$ and $\deg(g_2(x)) > 0$. Using the above exercise we deduce that $g_1(a_i) = g_2(a_i) = 1$ for all i, or $g_1(a_i) = g_2(a_i) = -1$ for all i. We may assume that $g_1(a_i) = g_2(a_i) = 1$ for all i. If $\deg(g_1(x)) < n$, we deduce that $g_1(x) = 1$ which is impossible. Similarly the case that $\deg(g_2(x)) < n$ does not occur. Now we come to the case that $\deg(g_1(x)) = \deg(g_2(x)) = n$. We see that $g_1(x) = b_1 f(x) + 1$ and $g_2(x) = b_2 f(x) + 1$ for some $b_1, b_2 \in \mathbb{Z}$. You see the contradictions.

$\square$

Chapter 3.

(7) **Answer.** $B \sim \begin{bmatrix} 1 & 0 & 0 & 0 \\ 0 & 3 & 0 & 0 \\ 0 & 0 & 21 & 0 \\ 0 & 0 & 0 & 0 \end{bmatrix}$.

(6) **Answer.** $2, 2, 156$.

(8) **Answer.** $\begin{bmatrix} 1 & 0 & 0 \\ 0 & x-1 & 0 \\ 0 & 0 & (x-1)(x-2) \end{bmatrix}$.

(9) **Answer.** $1, 1, 2x^2 + 3x$.

(15) **Proof.** Suppose that $\tilde{\varphi}_1, \tilde{\varphi}_2$ are two such extensions. Consider the module homomorphism $\tilde{\varphi}_1 - \tilde{\varphi}_2 : M \to N$. We see that $(\tilde{\varphi}_1 - \tilde{\varphi}_2)(S) = 0$. So $S \subseteq \ker(\tilde{\varphi}_1 - \tilde{\varphi}_2) \leq M$. Then $M = \langle S \rangle \subseteq \ker(\tilde{\varphi}_1 - \tilde{\varphi}_2)$ too. So $\tilde{\varphi}_1 - \tilde{\varphi}_2 = 0$, i.e., $\tilde{\varphi}_1 = \tilde{\varphi}_2$. Therefore $\tilde{\varphi} : M \to N$ is uniquely determined by the map φ. □

(18) **Proof.** (a). The statement that f is one-to-one simply says $C \cap D = 0$, and the statement that g is one-to-one says the same thing. Suppose f is onto.

(b). To prove g onto we take $e \in E$. Write $e = b + c$ according to the decomposition $B \oplus C$. Since f is onto, there exists $d \in D$ with $f(d) = b$. Write the decomposition of d as $d = b + c_1$. Then $c - c_1 = -d + e$, so that $g(c - c_1) = e$ as required. □

(22) Hints: Show that every maximal ideal in $\mathbb{Z}[x]$ is of the form $(p, f(x))$ where p is prime integer and $f(x)$ is primitive integer polynomial that is irreducible modulo p.

Chapter 4.

(12) **Solution.** Let $q = p^n$ where p is a prime. The polynomial $x^2 + x + 1$ reducible over F, iff there is $a \in F$ such that $a^2 + a + 1 = 0$. This is true for $q = 3$. Now we assume that $q \neq 3$.

Then the polynomial $x^2 + x + 1$ reducible over F, iff there is $a \in F^* \setminus \{1\}$ such that $a^3 - 1 = 0$, iff $3 | \phi(q) = p^{n-1}(p-1)$ since $(F^*, \cdot)$ is cyclic of order $\phi(q)$, iff $q = 3^n$, or $q = p^n$ with $3 | p - 1$. □

(15) **Proof.** Let F be a finite field of p^n elements. Let m be a divisor of n, so that $n = mq$ for some $q \in \mathbb{N}$. Then we know that the equation $x^{p^n - 1} = 1$ has exactly $p^n - 1$ distinct zeros in F. Let

$p^n - 1 = (p^m - 1)k$, we see that $k \equiv 1 \pmod{p}$. Then we know that

$$x^{p^n-1} - 1 = (x^{p^m-1})^k - 1 = (x^{p^m-1} - 1)(1 + x^{p^m-1} + \cdots + x^{(p^m-1)(k-1)})$$

for all $a \in F$. Since $\gcd(x^{p^m-1} - 1, 1 + x^{p^m-1} + \cdots + x^{(p^m-1)(k-1)}) = 1$, then there are exactly $p^m - 1$ zeros of $x^{p^m-1} - 1$ in F, i.e., all the p^m zeros of $x^{p^m} - x$ in F form a field of order p^m. This is the only such field since any such field satisfies $x^{p^m} - x = 0$. □

(21) **Solution.** Suppose $K \leq F$. Then $\mathbb{Z}_{13} \leq K \leq F$, and we have $11 = [F : \mathbb{Z}_{13}] = [F : K][K : \mathbb{Z}_{13}]$. We see that $[K : \mathbb{Z}_{13}]|11$. Then $[K : \mathbb{Z}_{13}] = 1$ or 11. There are exactly two subfield of F: F and $\mathbb{Z}_{13}$. □

(22) **Solution.** Let $r = \deg(\alpha, \mathbb{Z}_{13})$, $K = \mathbb{Z}_{13}(\alpha) \leq F$. Then $r = [K : \mathbb{Z}_{13}]$. From $11 = [F : \mathbb{Z}_{13}] = [F : K][K : \mathbb{Z}_{13}] = r[F : K]$, we see that $\deg(\alpha, \mathbb{Z}_{13})|11$. Then $\deg(\alpha, \mathbb{Z}_{13}) = 1$ or 11. □

(23) **Solution.** Let $f(x)$ be an irreducible polynomial of degree 11 over $\mathbb{Z}_{13}$ and E be the splitting field of $f(x)$. Since there is only one field $F_{13^{11}}$ of order 13^{11} we know that $[E : \mathbb{Z}_{13}] = 11$ and $E \simeq F_{13^{11}}$. Thus all 11 distinct zeros of $f(x)$ are in E. There are $13^{11} - 13$ elements in E that has degree 11 over $\mathbb{Z}_{13}$. So the number of distinct irreducible polynomials of degree 11 over $\mathbb{Z}_{13}$ is

$$\frac{13^{11} - 13}{11}.$$

□

(24) **Proof.** We know that $F = F_{p^n}$. Let

$$g(x) = \prod_{d|n} \prod_{j=1}^{r_d} f_{d,j}(x).$$

We will show that $x^{p^n} - x = g(x)$. It is clear that if $\mathbb{Z}_p \leq K \leq F$ then $|K| = p^d$ for some $d|n$. Each $\alpha \in F$ is algebraic over $\mathbb{Z}_p$ and has degree d that divides n. Thus $\operatorname{irr}(\alpha, \mathbb{Z}_p) = f_{d,j}(x)$ for a unique $d|n$ and for a unique $j = 1, 2, \cdots, r_d$, and each zero of $f_{d,j}(x)$ can generate the same extension field of $\mathbb{Z}_p$ since they have the same degree over $\mathbb{Z}_p$. Note that $f_{d,j}(x)|x^{p^n} - x$. Since F is finite, $f_{d,j}(x)$ has exactly d distinct zeros in F. So each $\alpha \in F$ has a unique $f_{d,j}(x)$ as its irreducible polynomial. Also each $f_{d,j}(x)$ has exactly d zeros in F. We see that F consists of all zeros of $g(x)$. Therefore $x^{p^n} - x = g(x)$. □

(30) **Answer.** $\operatorname{irr}(e^2 - 2, \mathbb{Q}(e^3)) = x^3 + 6x^3 + 12x + 8 - e^6$.

(32) **Proof.** $(a^2 + b^2)(c^2 + d^2) = N(a + bi)N(c + di) = N((a + bi)(c + di)) = N(ac - bd) + (ad + bc)i) = (ac - bd)^2 + (ad + bc)^2.$ □

(33) **Solution.** Suppose p is not irreducible in $\mathbb{Z}[i]$, and $p = (a + bi)(c + di)$ where $a, b, c, d \in \mathbb{Z}$, and both $a + bi$ and $c + di$ are not units. Then $p^2 = (a^2 + b^2)(c^2 + d^2)$. Since $(a^2 + b^2) \neq 1 \neq (c^2 + d^2)$, we know that $p = a^2 + b^2 = c^2 + d^2$ which is not true since $p \equiv 3 \pmod 4$. So p is irreducible in $\mathbb{Z}[i]$. □

(36) **Solution.** Note that $1146600 = 13 * 2^3 * 3^2 * 5^2 * 7^2$. By Theorem 4.4.10 we can write the integer as a sum of two integer squares. Actually

$$1146600 = 210^2 + 1050^2.$$

□

(38) **Proof.** If π were algebraic, πi would be algebraic as well. By Theorem 4.3.13, then $e^{\pi i} = -1$ would be transcendental, which is a contradiction. Therefore π is not algebraic, which means that it is transcendental. □

Chapter 5.

(2) **Solution.**

(a). We have the factorization

$$x^4 + 4 = (x^2 + 2x + 2)(x^2 - 2x + 2),$$

where the factors are irreducible by Schönemann-Eisenstein Criterion ($p = 2$). The roots are $\pm 1 \pm i$, so the splitting field is $\mathbb{Q}(i)$, which has degree 2 over $\mathbb{Q}$.

An alternate solution is to solve $x^4 = -4$. To find one root, use DeMoivre's theorem to get $\sqrt[4]{-4} = \pm 1 \pm i$.

(b). So the splitting field is $\mathbb{Q}(i)$, which has degree 2 over $\mathbb{Q}$.

(c). The Galois group $\mathrm{Gal}(\mathbb{Q}(i)/\mathbb{Q})$ must be cyclic of order 2, which is generated by the conjugation automorphism

$$\phi : \mathbb{Q}(i) \to \mathbb{Q}(i), a + bi \mapsto a - bi, \forall a, b \in \mathbb{Q}.$$

□

(3) **Solution.** Since $F = K(u, v) \supseteq K(u) \supseteq K$, where $[K(u) : K] = m$ and $[K(u, v) : K(u)] \leq n$, we have $[F : K] \leq mn$. But $[K(u) : K] = m$ and $[K(v) : K] = n$ are divisors of $[F : K]$, and since $\gcd(m; n) = 1$, we must have $mn | [F : K]$. So $[F : K] = mn$. □

(4) **Solution.** Since $u^2 \in K(u)$, we have $K(u) \supseteq K(u^2) \supseteq K$. Suppose that $u \notin K(u^2)$. Then $x^2 - u^2$ is irreducible over $K(u^2)$ since it has no roots in $K(u^2)$, so u is a root of the irreducible polynomial $x^2 - u^2$ over $K(u^2)$. Thus $[K(u) : K(u^2)] = 2$, and therefore 2 is a factor of $[K(u) : K]$. This contracts the assumption that $[K(u) : K]$ is odd. So $u \notin K(u^2)$ and hence $K(u^2) = K(u)$. □

(5) **Solution.** It is clear that the polynomial $x^3 - 11$ is irreducible over the field $\mathbb{Q}$. The roots of the polynomial are $\sqrt[3]{11}$, $a\sqrt[3]{11}$ and $a^2\sqrt[3]{11}$, where a is a primitive cube root of unity. Since a is not real, it cannot belong to $\mathbb{Q}(\sqrt[3]{11})$. Since a is a root of the irreducible polynomial $x^2 + x + 1$ over $\mathbb{Q}(\sqrt[3]{11})$ and $F = \mathbb{Q}(\sqrt[3]{11}, a)$, we have $[F : \mathbb{Q}] = [F : \mathbb{Q}(\sqrt[3]{11})][\mathbb{Q}(\sqrt[3]{11}) : \mathbb{Q}] = 2 \cdot 3 = 6$. □

(6) **Solution.** We know that $\mathrm{Gal}(F/\mathbb{Q})$ has odd order. If u is a nonreal root of $f(x)$, then since $f(x)$ has rational coefficients, its conjugate $\bar{u}$ must also be a root of $f(x)$. It follows that F is closed under taking complex conjugates. Since complex conjugation defines an automorphism of the complex numbers, it follows that restricting the automorphism to F defines a homomorphism from F into F. Because F has finite degree over $\mathbb{Q}$, the homomorphism must be onto as well as one-to-one. Thus complex conjugation defines an element of the Galois group of order 2, and this contradicts the fact that the Galois group has odd order. We conclude that every root of $f(x)$ must be real. □

(18) **Solution.**

(a). We have $x^8 - 1 = (x^4 - 1)(x^4 + 1) = (x - 1)(x + 1)(x^2 + 1)(x^4 + 1)$, giving the factorization over $\mathbb{Q}$. The factor $\Phi_8(x) = x^4 + 1$ is irreducible over $\mathbb{Q}$. The roots of $x^4 + 1$ are thus the primitive 8th roots of unity, $\frac{\pm 1 \pm i}{\sqrt{2}}$, and adjoining one of these roots also gives the others, together with i. Thus the splitting field is obtained in one step, by adjoining one root of $x^4 + 1$, so its degree over $\mathbb{Q}$ is 4.

It is clear that the splitting field can also be obtained by adjoining first $\sqrt{2}$ and then i, so $F = \mathbb{Q}(i, \sqrt{2})$.

(b). These subfields of $\mathbb{Q}(\sqrt{2}, i)$ are the splitting fields of $x^2 - 2, x^2 + 1$, and $x^2 + 2$, respectively. Any automorphism must take roots to roots, so if θ is an automorphism of $\mathbb{Q}(\sqrt{2}, i)$, we must have $\theta(\sqrt{2}) = \pm\sqrt{2}$, and $\theta(i) = \pm i$. These possibilities must in fact define 4 automorphisms of the splitting field.

Since all these automorphisms are of order 2, so the Galois group is $\mathbb{Z}_2 \times \mathbb{Z}_2$.

$\square$

Chapter 6.

(7) **Proof.** It is clear that $F \leq F(x^2) \cap F(x^2 - x)$. We need to show that $F(x^2) \cap F(x^2 - x) \leq F$. To the contrary, suppose not. Then there is a positive degree $f(x) \in F(x^2) \cap F(x^2 - x)$. We have the subfields

$$F \leq F(f(x)) \leq F(x^2) \cap F(x^2 - x) \leq F(x).$$

We see that $F(x)$ is a finite algebraic extension of $F(f(x))$. So any automorphism in the group $\mathrm{Gal}(F(x)/F(f(x)))$ is of finite order. Consider the following automorphisms $\tau_1 \in \mathrm{Gal}(F(x)/F(x^2))$, $\tau_2 \in \mathrm{Gal}(F(x)/F(x^2 - x)$ defined by

$$\tau_1(x) = -x, \tau_2(x) = 1 - x.$$

Clearly $\tau_1\tau_2 \in \mathrm{Gal}(F(x)/F(f(x)))$. But $\tau_1\tau_2(x) = x + 1$. Consequently, $\tau_1\tau_2$ is of infinite order, which is impossible. So $F(x^2) \cap F(x^2 - x) = F$. $\square$

(9) **Answer.** $\mathrm{irr}(\alpha + \alpha^2, \mathbb{Q}) = x^3 - 15x - 20$.

(16) **Solution.** Since F has characteristic zero, we know that F is a normal extension of $\mathbb{Q}$. So we can use Galois theorem. Because $\mathrm{Gal}(F/\mathbb{Q})$ is abelian, every subgroup is normal, and every intermediate extension between $\mathbb{Q}$ and F must be normal. Therefore if we adjoin to $\mathbb{Q}$ any root u of $f(x)$, the extension $\mathbb{Q}(u)$ must contain all other roots of $f(x)$, since it is normal over $\mathbb{Q}$. Thus $\mathbb{Q}(u)$ is a splitting field for $f(x)$ over $\mathbb{Q}$, so $\mathbb{Q}(u) = F$. $\square$

(18) **Solution.** Let G be the Galois group of $x^5 - 2$, and let ω be a primitive 5th root of unity. Then the roots of $x^5 - 2$ are $b = \sqrt[5]{2}$ and $\omega^j b$, for $1 \leq j \leq 4$. The splitting field of $x^5 - 2$ over $\mathbb{Q}$ is $F = \mathbb{Q}(\omega, b)$. Since $p(x) = x^5 - 2$ is irreducible over $\mathbb{Q}$ by Schönemann-Eisenstein Criterion, it is the irreducible polynomial of b. The element ω is a root of $x^5 - 1 = (x - 1)(x^4 + x^3 + x^2 + x + 1)$, so $\mathrm{irr}(\omega, \mathbb{Q}) = x^4 + x^3 + x^2 + x + 1$. Thus $[F : \mathbb{Q}] \leq 20$. Since

$$[\mathbb{Q}(\omega, b) : \mathbb{Q}(\omega)] = 4, [\mathbb{Q}(\omega, b) : \mathbb{Q}(b)] = 5,$$

the degree $[F : \mathbb{Q}]$ must be divisible by 5 and 4. It follows that $[F : \mathbb{Q}] \geq 20$. Therefore $|G| = 20$. $\square$

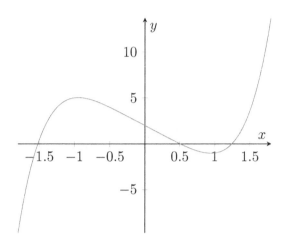

FIGURE 2. $y = x^5 - 4x + 2$

(19) **Solution.** The polynomial $f(x)$ is irreducible over $\mathbb{Q}$ since it satisfies Schönemann-Eisenstein Criterion for $p = 2$. Consider $y = f(x)$ as a continuous real valued function. The derivative $f'(x) = 5x^4 - 4$ has two real roots $\pm\sqrt[4]{\frac{4}{5}}$. We see that $f'(x) > 0$ if $|x| > \sqrt[4]{\frac{4}{5}}$, and $f'(x) < 0$ if $|x| < \sqrt[4]{\frac{4}{5}}$. That is, $f(x)$ is strictly increasing for $|x| > \sqrt[4]{\frac{4}{5}}$, and $f(x)$ is strictly decreasing for $|x| < \sqrt[4]{\frac{4}{5}}$. Since $f(-\infty) = -\infty, f\left(-\sqrt[4]{\frac{4}{5}}\right) > f(-1) = 5 > 0, f\left(\sqrt[4]{\frac{4}{5}}\right) < f(1) = -1 < 0, f(\infty) = \infty$, i.e.,

x	$-\infty$	$\nearrow$	$-\sqrt[4]{\frac{4}{5}}$	$\nearrow$	$\sqrt[4]{\frac{4}{5}}$	$\nearrow$	∞
$f'(x)$	∞	> 0	0	< 0	0	> 0	∞
$f(x)$	$-\infty$	$\nearrow$	> 0	$\searrow$	< 0	$\nearrow$	∞

we see that $f(x)$ must have exactly three real roots. It follows from a theorem that the Galois group of $f(x)$ over $\mathbb{Q}$ is S_5 which is not solvable. Thus $f(x)$ is not solvable by radicals. See Figure 2.   □

(36) Hints: Show that $\mathrm{Gal}(\Phi_{12}(x)/\mathbb{Z}_p)$ is not cyclic, i.e., the unit group $U(\mathbb{Z}_{12}, +, \cdot)$ is not cyclic.

Appendix A. Equivalence Relations and Kuratowski-Zorn Lemma

In this appendix we mainly recall some concepts and results from Set Theory, see [L] for details.

Definition A.0.1. *Let A, B be nonempty sets.* **The Cartesian product** *of A and B is the set*

$$A \times B = \{(a, b) : a \in A, b \in B\}.$$

Definition A.0.2. *Let S be a nonempty set. A* **binary relation** *(or a* **relation***) R in S is a subset of $S \times S$. We usually write aRb if $(a, b) \in R$.*

Definition A.0.3. *Let R be a relation in a nonempty set S. Then*

(1) R is called **reflexive** *if $(x, x) \in R$ for all $x \in S$;*

(2) R is called **symmetric** *if $(x, y) \in R \Rightarrow (y, x) \in R$;*

(3) R is called **anti-symmetric** *if $(x, y) \in R$ and $(y, x) \in R \Rightarrow x = y$;*

(4) R is called **transitive** *if $(x, y) \in R$ and $(y, z) \in G \Rightarrow (x, z) \in R$.*

Definition A.0.4. *A relation in a nonempty set S is called an* **equivalence** *relation if it is reflexive, symmetric, and transitive.*

Let R be an equivalence relation in a nonempty set S. If $(x, y) \in R$, we will write simply $x \sim y$ or $x \equiv y \pmod{R}$ and say that x is equivalent to y. Noting that R is an equivalence relation in S, for any $x, y, z \in S$ we have

(1) $x \sim x$,

(2) $x \sim y \Rightarrow y \sim x$,

(3) $x \sim y$ and $y \sim z \Rightarrow x \sim z$.

Definition A.0.5. *Let S be a nonempty set and let R be an equivalence relation in S. If $x \in S$, then the* **equivalence class of x modulo** *R is defined as follows:*

$$[x]_R = \{y \in S : y \sim x\}.$$

The collection of all the equivalence classes modulo R:

$$S/R = \{[x]_R : x \in S\}$$

is call the **quotient set** *of S modulo R.*

Definition A.0.6.   *(1) A nonempty set S is said to be* **partially ordered** *if a given binary relation $\leq$ in S satisfies:*
   (a) $a \leq a$, for any $a \in S$ (reflexive law),
   (b) $a \leq b$, $b \leq c \Rightarrow a \leq c$, for any $a, b, c \in S$ (transitive law),
   (c) $a \leq b$, $b \leq a \Rightarrow a = b$, for any $a, b \in S$ (antisymmetric law).
   (2) A partially ordered set S is said to be **totally ordered** *if for every pair $a, b \in S$ we have either $a \leq b$ or $b \leq a$.*
   (3) Let S be a partially ordered set. An elements $x \in S$ is called a **maximal element** *if $x \leq y$ with $y \in S \Rightarrow x = y$. Similarly, we can define a* **minimal** *element of S.*
   (4) Let T be a totally ordered subset of a partially ordered set S. We say that T has an **upper bound** *in S if there exists $c \in S$ such that $x \leq c$ for all $x \in T$.*
   (5) A totally ordered set S is **well-ordered** *if for every nonempty subset $X \subseteq S$, there exists $x \in X$ satisfying $y \geq x$ for all $y \in X$.*

Theorem A.0.7 (Kuratowski-Zorn Lemma). *Let S be a partially ordered set. If every totally ordered subset of S has an upper bound in S, then S contains a maximal element.*

Kuratowski-Zorn Lemma is also known as Zorn's Lemma. It was proved by Kazimierz Kuratowski (1896–1980) in 1922 and independently by Max Zorn (1906–1993) in 1935. Kuratowski-Zorn lemma is widely used in many situations.

Theorem A.0.8 (The Well Ordering Principle). *Any nonempty set S can be well-ordered, that is, there is a well-ordering on S.*

Theorem A.0.9 (The Axiom of Choice). *Given a class of nonempty sets, there exists a "choice function", i.e., a function which assigns to each of these sets one of its elements.*

In Set Theory, Axiom of Choice is logically equivalent to Kuratowski-Zorn Lemma which is logically equivalent to the Well Ordering Principle.

REFERENCES

[DF]   David S. Dummit, Richard M. Foote, *Abstract Algebra*, 3rd edition, Dummit and Foote, Wiley, 2003.

[F]   John B. Fraleigh, *A First Course in Abstract Algebra*, 7th edition, Addison-Wesley, 2003.

[J]   Nathan Jacobson, *Basic algebra*. I. 2nd edition. W. H. Freeman and Company, New York, 1985. xviii+499 pp.

[LZ]   Libin Li, Kaiming Zhao, *Introduction to Abstract Algebra*, ISBN: 978-7-03-067958-1, Academic Press, 2021.

[L]   Seymour Lipschutz, *Set Theory and Related Topics*, 2nd edition, McGraw Hill, 1998.

[M]   James S. Milne, *Fields and Galois Theory*, https://www.jmilne.org/math/CourseNotes/FT421.pdf, 2013.

[P]   Victor V. Prasolov, *Polynomials*, Translated from the 2001 Russian second edition by Dimitry Leites. Algorithms and Computation in Mathematics, 11. Springer-Verlag, Berlin, 2004.

[Z]   Kaiming Zhao, *Linear Algebra*, ISBN: 978-1-7924-6399-0, Kendall Hunt Publishing Company, 2021.

[ZTL] Kaiming Zhao, Haijun Tan, Genqiang Liu, *Group Theory*, ISBN: 978-1-7924-7892-5, Kendall Hunt Publishing Company, 2021.

Index

Printed in the United States
by Baker & Taylor Publisher Services